Technikzukünfte, Wissenschaft und Gesellschaft / Futures of Technology, Science and Society

Reihe herausgegeben von
A. Grunwald, Karlsruhe, Deutschland
R. Heil, Karlsruhe, Deutschland
C. Coenen, Karlsruhe, Deutschland

Diese interdisziplinäre Buchreihe ist Technikzukünften in ihren wissenschaftlichen und gesellschaftlichen Kontexten gewidmet. Der Plural „Zukünfte" ist dabei Programm. Denn erstens wird ein breites Spektrum wissenschaftlich-technischer Entwicklungen beleuchtet, und zweitens sind Debatten zu Technowissenschaften wie u.a. den Bio-, Informations-, Nano- und Neurotechnologien oder der Robotik durch eine Vielzahl von Perspektiven und Interessen bestimmt. Diese Zukünfte beeinflussen einerseits den Verlauf des Fortschritts, seine Ergebnisse und Folgen, z. B. durch Ausgestaltung der wissenschaftlichen Agenda. Andererseits sind wissenschaftlich-technische Neuerungen Anlass, neue Zukünfte mit anderen gesellschaftlichen Implikationen auszudenken. Diese Wechselseitigkeit reflektierend, befasst sich die Reihe vorrangig mit der sozialen und kulturellen Prägung von Naturwissenschaft und Technik, der verantwortlichen Gestaltung ihrer Ergebnisse in der Gesellschaft sowie mit den Auswirkungen auf unsere Bilder vom Menschen.

This interdisciplinary series of books is devoted to technology futures in their scientific and societal contexts. The use of the plural "futures" is by no means accidental: firstly, light is to be shed on a broad spectrum of developments in science and technology; secondly, debates on technoscientific fields such as biotechnology, information technology, nanotechnology, neurotechnology and robotics are influenced by a multitude of viewpoints and interests. On the one hand, these futures have an impact on the way advances are made, as well as on their results and consequences, for example by shaping the scientific agenda. On the other hand, scientific and technological innovations offer an opportunity to conceive of new futures with different implications for society. Reflecting this reciprocity, the series concentrates primarily on the way in which science and technology are influenced social and culturally, on how their results can be shaped in a responsible manner in society, and on the way they affect our images of humankind.

Weitere Bände in der Reihe http://www.springer.com/series/13596

Matthias Braun · Hannah Schickl
Peter Dabrock
Editors

Between Moral Hazard and Legal Uncertainty

Ethical, Legal and Societal Challenges of Human Genome Editing

Springer VS

Editors
Matthias Braun
Chair of Systematic Theology II (Ethics)
Friedrich-Alexander-Universität
Erlangen, Germany

Peter Dabrock
Chair of Systematic Theology II (Ethics)
Friedrich-Alexander-Universität
Erlangen, Germany

Hannah Schickl
Interdisciplinary Research Group
"Gene Technology Report"
Berlin-Brandenburg Academy of
Sciences and Humanities
Berlin, Germany

ISSN 2524-3764 ISSN 2524-3772 (electronic)
Technikzukünfte, Wissenschaft und Gesellschaft / Futures of Technology, Science and Society
ISBN 978-3-658-22659-6 ISBN 978-3-658-22660-2 (eBook)
https://doi.org/10.1007/978-3-658-22660-2

Library of Congress Control Number: 2018962982

Springer VS

This Springer VS imprint is published by the registered company Springer Fachmedien Wiesbaden GmbH part of Springer Nature
The registered company address is: Abraham-Lincoln-Str. 46, 65189 Wiesbaden, Germany

Contents

Between Moral Hazard and Legal Uncertainty: An Introduction 1
Matthias Braun, Hannah Schickl and Peter Dabrock

I Mapping the Current Legal Framingg

Genome Editing and the Law. Some Remarks on Current Legal Challenges of CRISPR-Cas9 17
Stephan Rixen

Integrating Ethical Standards into the Human Rights Framework. Considerations towards the Future Regulation of Genome Editing on an International Level 31
Fruzsina Molnár-Gábor

II Theoretical Challenges

Prospects of Human Germline Modification by CRISPR-Cas9 – an Ethicist's View 53
Dieter Birnbacher

Genome Editing, Non-Identity and the Notion of Harm 67
Alina Omerbasic

Genome Editing and the Transgression of the Species Boundary.
Does Species Affiliation Have an Ethical Relevance? 83
Markus Rothhaar

Germline Modifications as a Severe Intervention into Human Nature 99
Nadia Primc

Is Genome Editing Unnatural? Nature in Bioethics, the Politics
of Bioethics, and the Political Dimension of Nature 111
Eva Odzuck

Transition and Care. Theological Concepts of Dynamic Creation
and the Ethics of Genome Editing 129
Mathias Wirth

Human Germline Genome Editing in the Clinical Context.
The Case of Disease Prevention 149
Giovanni Rubeis

III Towards a Governance Frame in Dealing with Genome Editing

Who? What? How? Why? If You Don't Ask You'll Never Know.
On Criticism of the New Uproar about Germline Editing –
Discourse Analytical and Socioethical Metaperspectives 163
Peter Dabrock

Debating Genome Editing Technologies. Questioning the Role
of Public Discussion 187
Lukas Kaelin

A Path Through the (Moral) Morass. Genome Editing, Reproduction
and Broad Conversations 203
Giulia Cavaliere

Authors

Dieter Birnbacher, Univ.-Prof. i. R. Dr. Dr. h.c., is Professor Emeritus of Practical Philosophy at the Heinrich-Heine University Düsseldorf and President of the German Society for Dying with Dignity. His main fields of research are concerned with ethical and anthropological issues in modern medicine as well as anthropology and neurophilosophy, e.g. theories of emotions.

Matthias Braun, Dr., is Assistant Professor at the Chair of Systematic Theology II (Ethics), Friedrich-Alexander University Erlangen-Nuremberg. His research is focused on theological, philosophical and socio-political concepts in dealing with emerging biotechnologies at the science-society interface (with special regard to genome editing, synthetic biology, stem cell research and big-data medicine) on the one hand as well as concepts of recognition and vulnerability and its implications for ethics on the other hand.

Giulia Cavaliere is a PhD student at the Department of Global Health & Social Medicine, King's College London. Her research interests concern philosophical and sociological approaches to bioethics, particularly the intersection between moral and social questions surrounding reproductive genetic technologies and the history of eugenics.

Peter Dabrock, Prof. Dr., is Professor of Systematic Theology (Ethics) at the Friedrich-Alexander University Erlangen-Nuremberg and Chair of the German Ethics Council. From 2011 until 2016 he was also member of the "European Group on Ethics in Science and New Technologies". His research focuses on concrete ethics from a Protestant perspective, foundational theology and bioethics as theologically based social ethics (with special regard to healthcare, human genomics, synthetic biology, big data and biobanks).

Lukas Kaelin, Dr., is Assistant Professor at the Institute of Practical Philosophy and Ethics, Catholic Private University of Linz. His research interests encompass social and political philosophy (especially the notion of the ‚public sphere') as well as bioethics and medical ethics.

Fruzsina Molnár-Gábor, Dr., is group leader at the Heidelberg Academy of Sciences and Humanities and at the BioQuant Centre of the University of Heidelberg, as well as lecturer at the Faculty of Law at the University of Heidelberg. Her research focuses on the regulation of biomedicine, including the legal fields of public international law, the law of the European Union, and comparative public law.

Eva Odzuck, Dr., is Assistant Professor at the Chair of Political Science II, Friedrich-Alexander University Erlangen-Nuremberg. Her research is focused on political philosophy, political theory, and history of political thought, as well as ethics and politics of the life sciences.

Alina Omerbasic, M.A., is Research Assistant at the Chair of Applied Ethics, Department of Philosophy, University of Potsdam. Her main fields of research are applied ethics and political philosophy.

Nadia Primc, Dr., is Research Assistant at the Institute of History and Ethics of Medicine, Medical Department, University of Heidelberg. Her research is concerned with philosophy of science as well as bio-, medical and nursing ethics and moral philosophy.

Stephan Rixen, Prof. Dr., is Professor at the Chair of Public Law I, Public Law, Social Economy Law and Health Law, University of Bayreuth. His research is focused on constitutional and administrative law (especially public commercial law), German and European social law (including children and youth welfare law) as well as health law.

Markus Rothhaar, Prof. Dr., is Endowed Chair for Bioethics at the Catholic University of Eichstätt-Ingolstadt. His research is focused on medicine and bioethics, especially the status of human embryos, bioethics and political philosophy, notions of life and justice in public health service.

Giovanni Rubeis, Dr., is Research Assistant at the Institute of the History, Philosophy and Ethics of Medicine, University of Ulm. His research focuses mainly on ethics in prenatal and reproductive medicine, philosophy of science of biology and theory of medicine.

Hannah Schickl, State Examination, is currently Research Assistant at the Interdisciplinary Research Group "Gene Technology Report", Berlin-Brandenburg Academy of Sciences and Humanities, Berlin, and PhD student at the Institute for Ethics, History and Theory of Medicine, University of Münster. Her research is focused on analytic bioethics, particularly gene technologies like stem cell research, reproductive medicine and animal ethics.

Mathias Wirth, Dr., is an Assistant Professor (tenure track) of Systematic Theology/Ethics at the School of Theology at the University of Bern (Switzerland). Prior to this position, he was a Feodor Lynen Postdoc at the Yale Divinity School (USA) (2017-2018) and a research associate and lecturer at the Department of Medical History and Ethics at the University Medical Center Hamburg-Eppendorf (2012–2017). His research is on medical ethics, trans*phenomena and religious ethics.

Between Moral Hazard and Legal Uncertainty

An Introduction

Matthias Braun, Hannah Schickl and Peter Dabrock

1 The Debate on Human Germline Genome Editing

Genome editing techniques are seen to be at the frontier of current research in the field of emerging biotechnologies (The Academy of Medical Sciences et al. 2015; Leopoldina et al. 2015; BBAW 2015; Academy of Sciences Leopoldina 2017). However, such biotechnological research is tensioned at the interface of science, technology and society (Braun and Dabrock 2016; Braun and Dabrock 2017). On the one hand, this means that it offers a tremendous potential to provide new concepts, methods and – in the long run – novel applications for urgent challenges and needs within society. On the other hand, such techniques are also perceived as having the potential to challenge and sometimes blur existing attitudes, values as well as comprehensive cultural modes of public understanding. The latest revolutionary development within genome editing, the so-called CRISPR technologies, represents a paradigmatic example of the ambiguity of such techniques. In the case of CRISPR, this ambiguity has resulted in an international interdisciplinary debate on whether or not it is necessary to completely ban, or temporarily prohibit, the application of this technique by means of a moratorium on its use for germline modifications, particularly in human embryos in the reproduction process (Lanphier et al. 2015; Baltimore et al. 2015). Interestingly, it was mostly a broad alliance of scientists themselves who initially called for a moratorium and pointed out the need for a public debate about how to deal with the possible applications of this technology (Hurlbut 2015). Unlike previous debates on, for example, stem cell research, synthetic biology, green gene technology and other controversial gene technologies, there has been neither joint action of the different civil science organisations to initiate a public debate on genome editing, nor a call for a moratorium. Most of the civil society organisations who gave a statement on CRISPR at all, merely agreed with the previous scientific call for a moratorium (Center for Genetics and Society 2015).

M. Braun et al. (Eds.), *Between Moral Hazard and Legal Uncertainty*,
Technikzukünfte, Wissenschaft und Gesellschaft / Futures of Technology,
Science and Society, https://doi.org/10.1007/978-3-658-22660-2_1

Following this early call for a moratorium the professional debate among ethicists and lawyers focussed on possible ethical reasonings and legal regulations in dealing with genome editing techniques (Cyranowski and Reardon 2015). In parallel, the use of CRISPR technologies, particularly for the modification of the human germline in the context of reproductive medicine, became part of a broader public reasoning that posed the question of which argumentation should be adduced in order to develop a societal and, consequently, political framework for dealing with such technologies, their opportunities and risks (Hurlbut 2015). In the course of this, societal discontent is an indicator for occurring trans- and reconfigurations of traditional and hitherto taken-for-granted-distinctions, such as 'living' vs. 'non-living', 'human' vs. 'non-human' or 'artificial' vs. 'natural' (Dabrock et al. 2013). In general, studies on the public perception of genome editing techniques indicate an apparent, more or less global consensus that no human germline should undergo genetic modification (Ishii 2014). And yet, germline cells are not as such seen as something miraculous or as some kind of Holy Grail. Rather, their modification is viewed more as the crossing of a symbolic border – symbolic in the sense that germline modification seems to impair an understanding of what it means to be human, become human or give the opportunity to become human in a specific manner. Remarkably, however, this symbolic border has already been crossed, for example, by the permission and application of mitochondrial (mt) DNA transfer techniques – which also modify the germline – in Mexico and the UK in 2016. Thus, there are already genetically modified entities (Nature Editorial 2017) without there having been any calls for a moratorium with respect to these germline engineering techniques. Against this background, there is an apparent lack of clarity regarding the precise reference of social unease about the modification of the germline in human embryos, in the context of reproduction, by genome editing techniques. Inasmuch as the two mtDNA transfer techniques currently applied in clinics (pronuclear and spindle transfer) replace the nuclear DNA in unfertilised egg cells or pronuclei (fertilised egg cells before the nuclear fusion), one might assume that the unease has something to do with the modification of human embryos (and their possible transfer to a woman and subsequent birth).[1] This is just one example of implicit presuppositions or open questions in the current debate. Thus, the urgent questions with respect to the application of genome editing techniques in humans, particularly

1 Cf. Nuffield Council on Bioethics 2016, 115: "Of all the potential applications of genome editing [...] the one that has consistently generated most controversy is the genetic alteration of human embryos *in vitro* and the possibility that altered embryos could be transferred to a woman who would give birth to a human being with a unique, altered genome".

for germline modification, could be described as follows: What are the underlying normative presumptions, comprehensive doctrines or values which constitute the appraisal of such applications and the related formulation of fundamental ethical and legal conflict lines? These lines can be seen in particular between the possibility of eliminating a substantial risk of disease for following generations and potential damage to future persons (in the case of possible side effects), as well as a perhaps problematic narrowing of the future of hitherto not existing persons in conjunction with the non-reversibility of the conducted genetic modifications. Thus, the main problem is not the use of genome editing techniques itself. The main task is to reach a stage at which it is possible to outline the underlying moral and legal reasoning, in order to build a sustainable framework for ongoing (political) decision making at the interface of science and society. With this in mind, the aim of this anthology is to make a substantial contribution to the current debate about a responsible and participatory framework for research on emerging biotechnologies such as CRISPR, by analysing underlying perceptions, attitudes, arguments and the reasoning. In other words, the concern of this book is not to develop or provide a certain kind of statement – but instead to take time for solid and in-depth thinking about difficult legal, societal and ethical questions evoked by emerging biotechnologies.

2 The Technology and the Shift in the Current Debate

It is important to take into account that CRISPR (cluster regularly interspaced short palindromic repeat) is not reported to be a breakthrough technology in the sense that it would first and foremost enable genome editing. There are other techniques, such as TALENs (transcription activator-like effector nucleases) or ZFNs (zinc finger nucleases), which have been used to edit the genome of living cells in different organisms for more than 15 years (Gasiunas and Siksnys 2013). However, the hitherto used techniques such as TALENs or ZFNs are quite expensive, because they require protein engineering to target specific DNA-sequences. By contrast, CRISPR-Cas9 uses guiding RNA molecules to target the respective DNA-sequence. These RNA molecules are much cheaper and can be synthesised more easily (Dow 2015). Thus, by now, every standard molecular laboratory is – or at least will be in the very near future – able to edit genes or whole genomes (Caplan et al. 2015). Currently, CRISPR-Cas9 is already in use for the modification of animals (particularly insects and mice), plants and microorganisms, and aims – as one important field of application – at the development of new therapeutic approaches (Ledford 2015).

Following several rumours that genome editing methods will shortly be used to conduct germline modifications in the human germline, a Chinese group was the first that was able to publish corresponding results in March 2015 (Liang et al.). In 2017, two other groups published further results from working with genome editing techniques on human germline cells (Ma et al.) as well as in cloned primates (Liu et al. 2017). The aim of the use of genome editing techniques for germline modification is to correct genetic defects, which would cause diseases, by enabling high-precision insertion and deletion, as well as the recording of genes in human egg cells, sperm cells and embryos (Hurlbut 2015). Liang and colleagues used CRISPR technologies in human embryos, which were incapable of developing further in vivo, to modify a gene that is capable of causing a potentially fatal blood disorder. Such modification could potentially affect every cell in a resulting organism and would likely be passed on to the following generations. Despite the undeniable achievements of the Chinese group led by Junjiu Huang, it needs to be pointed out that, firstly, only 28 of the initial 86 embryos were successfully spliced, of which, finally, only four contained the designed genetic material (Liang et al. 2015). Secondly, the researchers reported a (from their perspective) surprising number of so-called 'off target' mutations. This gives a first hint that the inserted genetic sequences seem to interact with other parts of the genome in an – at present – unknown way. Thirdly, the rate of such unintended mutations was much higher than those observed in mouse models or adult human cells (Liang et al. 2015).

While there seems to be a consensus on biosafety issues regarding accidental releases of particularly modified microorganisms, as well as on the importance of an appropriate health risk assessment with respect to the medical application in humans, the ethical debate becomes especially controversial regarding the question of whether CRISPR technologies could, should or even must be used for germline modification, particularly in the context of reproduction (Musunuru 2017). There are only few current suggestions to ban this application completely (Lanphier et al. 2015). Most of the current comments demand or at least endorse a (voluntary) moratorium on the use of CRISPR technologies for germline modification (Baltimore et al. 2015; BBAW 2015; Academy of Sciences Leopoldina et al. 2015). Up to now there is no common international regulatory framework regarding the use of genome editing techniques for germline modifications. The Oviedo Convention (Convention on Human Rights and Biomedicine), the only international legally binding instrument on the protection of human rights in biomedicine, prohibits interventions in the human genome that aim at modifying not only the genome of one person but also of his descendants (art. 13). In Germany, the Embryo Protection Act (*Embryonenschutzgesetz*, ESchG), one of the most restrictive laws on assisted reproduction and biomedical research, already prohibits the alteration of the genetic information of a human germline cell – whether in reproduction or research (art. 5 para. 1 ESchG). In contrast, the Human Fertilisation and Embryology Authority (HFEA) in the UK may decide on assisted reproduction, and research with gametes and embryos in vitro, on a case by case basis. Based on this more open regulation, the UK has taken a leading role in the EU in the conducting genome editing techniques in human embryos. Similarly, the UK was the first country in the world whose legislation allowed the use of mtDNA transfer (Gross 2015). Therefore, while the experiments of Liang and colleagues on germline modification were conducted only with non-viable human embryos (and thus in the research context), where mtDNA transfer is concerned, the UK already carries out germline modifications as a clinical procedure within assisted reproduction. It does so based on a deliberative process, discussed by a democratically assigned parliament, and finally constituted in a legal framework. However, there is at this point an intense debate about whether techniques such as CRISPR could be integrated into the legal *de lege lata* frames, or whether new *de lege ferrenda* legal frames need to be developed at the national, international and global levels (Academy of Sciences Leopoldina 2017).

Future regulation frameworks have to be built first and foremost on reliable and precise distinctions. By way of example, the use of CRISPR technologies in humans requires clear divisions of the context: research and application; as well as of the object: embryos (in research or reproduction) or born humans (somatic therapy). While research using gene editing techniques to develop new therapeutic approaches seems to be more welcome (as the future application of somatic therapies in patients is mostly desired), the modification of human embryos, particularly in reproductive contexts but also in research contexts, and here especially in germline cells, is highly controversial (German Ethics Council 2017). However, the initial broad consensus on a moratorium – backed by the Oviedo Convention as well as by nearly all scientific and political institutions – has recently come under pressure in light of new developments (Ma et al. 2017) and institutional statements. In their latest statement the US National Academies of Science argue for an important shift of future regulation, by switching away from 'not allowed as long as the risks have not been clarified' to "allowed *if* the technical challenges are overcome and potential benefits are reasonable in light of the risks" (The National Academies 2017). This switch in reasoning may mean that the US academies are no longer advocating a partially fundamental, partially risk-related rejection of germline therapy by genome editing, but instead a general permission guided by formal and material criteria (German Ethics Council 2017). Later on, in autumn 2017, a huge alliance of several scientific institutions stated once again the necessity of a clear distinction between the different contexts of research and reproduction (The National Academies 2017). They argue that, given the nature and number of unanswered scientific, ethical and policy questions, it is inappropriate to perform germline genome editing that culminates in human pregnancy. Besides their plea for a 'red light' on its use in the context of reproduction, they argue that there is no reason – at least not with appropriate oversight and consent from donors – to prohibit in vitro germline genome editing on human embryos and gametes for the facilitation of research on possible future clinical applications of genome editing (Ormond et al. 2017).

3 Mapping the Underlying Ethical Arguments

With respect to human germline genome editing has been argued that it is our moral duty "to remove all disease-causing genes from an embryo as this will lower the total frequency of disease-causing genes in the gene pool, and therefore the incidence of such diseases in future generations" (Gyngell et al. 2017). In this line of thought, editing human germline cells is seen as a responsible form of enhancing (the future of) following generations. While most people would agree that it is good and necessary to find ways to cure diseases, not many would respond to a strong plea for trying to prevent and ultimately eradicate diseases across generations. The various objections against germline genome editing can basically be divided into 1) fundamental (i.e. against the context of research and application) and 2) non fundamental (i.e. only against the context of application) arguments. The most prevalent fundamental arguments are a) arguments of human dignity, b) arguments of naturalness and c) slippery slope arguments.

a) With regard to human germline genome editing, arguments from human dignity are raised in two main versions: The first version refers to the embryo's human dignity as a subjective basic right in the sense of its (and the future child's as well as the descendants') instrumentalization by modifying its/their genes (BBAW 2015; Academy of Sciences Leopoldina et al. 2015). This version relies on arguments from the human embryo's (moral and legal) status (so called SKIP arguments) and may also include notions of a 'right to an open future', or a 'right to self-determination', based respectively on the lack of autonomy and capacity for consent. The second version refers to the human dignity of human beings in the sense of an objective obligation of the state to protect human dignity. It is therefore based on an abstract image of humanity (*Menschenbild*), or the human species, and its identity as an intrinsic value (Joerden et al. 2013; Rothhaar 2015): First, the assumption that human embryos in vitro already have human dignity as born human beings – which also underlies the German Embryo Protection Act – is highly controversial (Academy of Sciences Leopoldina 2017). The potentiality argument in particular, which is considered to be the strongest argument for absolute embryo protection, is increasingly criticized by ethicists in recent years (Stier and Schoene-Seifert 2013; Schickl et al. 2014). But, even if one follows this strong assumption, it can be argued, on the one hand, that the genetic modification neither instrumentalizes the embryo (for other purposes), nor violates its human dignity, since it is a medical intervention for the sake of the embryo's health and life (and therefore for his own sake). On the other hand, there is no sense in which the procedure humiliates the embryo, and it is doubtful that the born child or adult will complain about having been born healthy instead of diseased, or never having been born at all. Therefore, the assumed violation of human dignity can only lie in the alteration of the human genome itself. However, due to genomic variation there is no such thing as a 'human genome' shared by all of humanity (The National Academies 2017). And if there were, it is unclear why it should not be altered in order to prevent diseases (BBAW 2015).

b) In the context of human germline editing arguments of naturalness refer to the general unavailability of the genetic constitution, including genetic diseases, due to their naturalness as an intrinsic value. A theological version of this type of argument can also refer to illegitimately 'playing god' by intervening in creation (Dabrock 2009; Peters 2014). First of all, inasmuch as these arguments do not also speak against the intervention in animal genetic constitution, arguments of naturalness are actually arguments from human dignity (in the above mentioned second abstract version) in disguise; at issue is *human nature*, not nature as such. Secondly, the underlying assumption of naturalness as something entirely good is highly doubtful (Birnbacher 2014). Thirdly, these arguments lead in consequence to rejecting *any* (particularly medical) intervention, which is taking it much to far, as few (even among the proponents of these arguments) would agree with such a claim.

c) Applied to human germline editing, slippery slope arguments mainly warn against the danger of eugenics (which leads to 'designer babies', enhancement or 'human breeding'), of an unequal access to therapies, or of the discrimination against sick and disabled people (BBAW 2015; Academy of Sciences Leopoldina et al. 2015; Lanphier et al. 2015). Yet such arguments do not criticize germline therapies themselves as ethically unacceptable. Rather, they criticize the ethically unacceptable (legal and social) consequences that are *supposed* to arise from the permission and application of such therapies. However, precisely this structure of slippery slope arguments is their greatest weakness: They remain suppositions. Of course, these feared developments are abstracted from actual developments or tendencies of other technologies (e.g. the expanded catalogue of – meanwhile also non medical – indications in favor of a preimplantation genetic diagnosis). At the same time, however, the same technologies show that they have still not led to the predicted, ethically unacceptable consequences. Ultimately, slippery slope arguments cannot convince of their implicit assumption, that legal or other barriers are not sufficient to control the development of technologies.

The most common argument within the ethical (as well as legal) debate on the use of genome editing techniques, like CRISPR technologies, is the safety argument as a non fundamental objection. Typically, it speaks only against application and not against the research context. It points out that the technique is (still) too unsafe for its application in humans. There is consensus on this view. Yet what follows from it is controversial. The argumentation often refers to alternative, already existing screening techniques, like preimplantation genetic diagnosis (PGD) or prenatal diagnosis (PND), pointing out that they can be used with more foresight, and thus with more safety, for the prevention of inherited genetic diseases. This safety argument has been so strongly emphasized, to the effect that it is considered to imagine a situation in which the use of genome editing in human germline cells would offer a therapeutic benefit over existing and developing methods, in view of the fact that "the precise effects of genetic modification to an embryo may be impossible to know until birth" (Lanphier et al. 2015). In response to this view, Julian Savulescu and colleagues argued that, in cases of single gene disorders, it is possible that there is a couple for whom PGD is not an option (Gyngell et al. 2017). An example of this would be a case where two carriers of the gene for cystic fibrosis wish to have descendants together. However, if they only produce one embryo with two copies of the cystic fibrosis gene, there would be no option to use a PGD, as there is no healthy embryo to choose. In other cases, like Huntington's disease, it would not even help if there were more than one embryo, since every embryo produced from the parental gametes would be predisposed to this disease (Gyngell et al. 2017). However, it can be objected that these cases in which PGD or PND are not an option (both parents are homozygous for a recessive disease or one is homozygous for a dominant disease) are in fact rare, and that therefore, for most of the cases the safety argument referring to alternative methods remains valid. On that basis, it can even be argued that, given the improbability of these 'theoretically construed cases', there is no medical necessity at all for germline therapies and, therefore, there is also no need to develop these therapies in research in the first place (Hyun and Osborn 2015). Besides, it can be argued that in the cases where PGD is an actual alternative its application is ethically *always* preferable to the application of genome editing techniques, as long as there remains even a little health risk (because this little health risk would still be bigger than the minimal to zero risk of PGD). In this version the safety argument becomes much stronger, viz. substantial – in contrast to the usual temporary meaning (it also can become fundamental in contrast to the typical non fundamental version) – thereby questioning the (meanwhile widespread, international) view that germline modifications should be allowed and conducted as soon as the risks are 'sufficiently safe' (whatever 'sufficiently' means here). Contrary to the often expressed criticism, that there are no new ethical arguments within the current debate on genome editing (Academy of Sciences Leopoldina 2017), this line of argumentation is actually new.

4 The Composition of this Anthology

The above presented ethical questions and arguments can be mapped around three – distinguished but not divided – perspectives. In the first section of this anthology, the addressed challenges are stressed from legal perspectives, in order to pave the way for and embed the second theoretical and conceptual section. The third section combines both perspectives by focussing on concrete challenges for a future governance of human genome editing.

The first section of this anthology is based on the idea, that in order to explore and map the field of ethical reasoning, it is crucial to gain a precise overview of the concrete legal frameworks regarding human germline editing, on the one hand, and the interdependency of legal considerations with underlying public perceptions, personal attitudes, moral values and ethical arguments on the other. Along these lines, the first chapter, written by Stephan Rixen, reflects on the function of law in biomedicine. Starting form such a foundational view on the possibilities and boundaries of law, he gives a comparative insight into the modes of legal regulation in Germany and the UK. He concludes that both legal systems are embedded in specific forms of cultural diversity: "If we accept that truth-driven approaches to biomedical problems are not compatible with the task of law in a pluralistic society, then we can imagine and implement a design of law which makes room for what is more important: the encounter of ethical diversity, and its transformation, with step by step regulation." In this line, German law could and should learn from the UK legal frame.

In addition to this legal comparison on a national level, and against the background that an internationally harmonized approach to the application of genome editing in humans is still missing, Fruzsina Molnár-Gábor analyzes the chances for and limits of a future UNESCO declaration on genome editing. In doing so, she focusses particularly on the integration of ethical standards into the human rights framework, the binding effect of such a possible future instrument, the role of deliberation during its development, and the possible long-term effects on the human rights system.

The second section of this anthology is concerned with the more theoretical challenges evoked by the upcoming (partly visionary) procedures of human genome and germline editing. For this, the authors analyze underlying ethical arguments – particularly, arguments of safety, naturalness and human dignity – as well as general assumptions within the ethical debate. This section starts with a foundational chapter on the ethical role and meaning of the safety argument in the debate on human germline editing. Dieter Birnbacher gives an overview of the different arguments against interventions in the human germline and points out that there is only one argument that is sufficiently strong to be practically relevant: The argument that it is doubtful whether the challenge of off-target effects, with potentially fatal health consequences for future generations, can be met. In line with the above outlined substantial (and also fundamental) version of the safety argument, Birnbacher concludes that there are good reasons to promote PGD rather than germline gene editing, as nearly all objectives of human germline gene editing can also be attained by PGD, and PGD is in comparison less risky.

From a broader perspective, Alina Omerbasic analyzes the general assumption (partially also underlying the safety argument) that the embryo and its descendants are harmed by germline editing. From an analytic-philosophical perspective, she questions whether the application of CRISPR technologies on humans violates the principle of nonmaleficence. Her two-tier analysis distinguishes between the question of harm to the edited embryo itself and harm to its descendants. Even though we face the so-called Non-Identity Problem in the latter case, which blocks the route to a harm done-result, she tries to set boundaries in an apparently harm-free area by introducing a new action-guiding principle for such cases.

By asking whether the possible modification of the human genome would lead to an erosion of species boundaries, Markus Rothhaar analyzes the species argument (related to the above mentioned argument of human dignity in its second version, and respectively, to the argument of naturalness). Taking this question seriously, he scrutinizes whether an eventual creation of transgenic organisms with both human and animal genes would put into question bioethical arguments that rely on species affiliation. After a close reexamination of such arguments, he concludes that the transgression of species boundaries on a merely genetic level does not undermine their validity. Possible epistemic uncertainties about the moral status of transgenic organisms nonetheless call for a precautionary approach.

In connection with this line of argumentation, Nadia Primc, Eva Odzuck and Mathias Wirth analyze the role of naturalness in the context of human genome editing from different perspectives. Nadia Primc analyzes whether human germline editing is – as sometimes criticized – a severe intervention into human nature. In order to assess the meaning and tenability of these claims, she differentiates between different uses of the term 'human nature' as well as distinct scientific and clinical contexts. She concludes that as an unproven intervention, germline manipulation must be regarded as a severe intervention into the nature of a prospective newborn – an assertion that no longer applies to the hypothetical case that germline manipulation can be regarded as a safe therapeutic option, as long as the intervention is restricted to the prevention of severe genetic diseases.

In a similar direction – but with the focus on the political implications – Eva Odzuck argues that a responsible policy of the human genome presupposes the spelling-out of the normative and political dimensions of nature. By drawing a line from early modern natural science, and its goal to dominate and control nature (understood in mechanical and materialistic terms), to modern liberal positions, and their readiness to intervene into the living human body of another person, she insists that our relation to nature and to the living human body is a deeply normative and political one that needs to be a focus of political theory research and in public discourse.

In line with the above mentioned theological version of the argument of naturalness, Mathias Wirth critically examines arguments which postulate that modifications of the human germline, and thus of a very fundamental dimension of the bodily sphere, seem in some way to contradict what theologically is considered to be a good creation. Against this background he aims to add affirmative arguments, from a Christian theological conception in the vicinity of process theology, to the debate on genome editing and to the transition of the biological basis of human beings.

Giovanni Rubeis examines the general ethical assumption that germline editing prevents diseases. He shows that only one possible application of germline editing, namely zygote editing, can be thought of as a means of disease prevention. Most applications have instead to be considered as advanced techniques of assisted reproduction. From that point of view, this renewed framing is important when it comes to the risk assessment of these applications as well as to questions of funding.

The third section of the anthology broadens the perspective towards concrete challenges within the governance of new genome editing technologies. Debates about moral, legal and political attributions of responsibility do not, as can be seen in the past, occur in a vacuum. With this in mind, Peter Dabrock explores the genealogy of the climate of the current debate by uncovering the transition from the at a first glance rather reserved statement of the Washington Summit of December

2015, to the about-face undertaken by the leading American academies (who were themselves involved in the Washington Summit) in the report they published in February 2017. Peter Dabrock critically scrutinizes the hidden genealogy of discourses in order to gain a more in-depth understanding of their societal framing. Making these taken-for-granted assumptions explicit is of intrinsic value for further ethical debate. Finally, Dabrock explains how trust in science can be maintained beyond such perceptible interest-led transformations.

Lukas Kaelin takes a closer look at the notion of 'the public' often referred to in ethical statements and position papers on genome editing, which claim that a public discussion should take place. As it is not self-evident what this reference to the public means and why such a public engagement should take place, he explores the concept of the public as discussed in key position papers on genome editing and puts it in relation with political theory. He points out that the notion of the public remains fuzzy, sometimes pertaining to collective problem-solving, sometimes to education and self-determination among citizens. An important aspect in the engagement of the public, he concludes, consists in creating an environment friendly to innovation and research.

Finally, Giulia Cavaliere discusses and appraises the potential strategies that can be adopted to address the moral disagreement (also present in the chapters of this anthology) that surrounds the use of genome editing technologies in the context of assisted reproduction. Reflecting on the opposing arguments within the debate, and the strategies used in bioethics in the event of moral disagreement, she contends that the clarification of the facts regarding a new technology, as well as the formulation of the best philosophical argument, might result in states of affairs that fail to promote peoples' well-being. Instead, she concludes, we should focus on finding at least some common ground in order to move forward in the debate and to develop morally sound policies, whilst still accommodating opposing views as much as possible.

The chapters of this book are based on a conference week pertaining to the topic: "Between Moral Hazard and Legal Uncertainty. Ethical, Legal and Societal Challenges in Dealing with Genome Editing". Both the conference and the preparation of this book have been funded by the German Ministry of Education and Research. The single chapters were written by the different authors based on their talks and the intensive discussions during the conference, and have been reviewed by the editors. We thank all participants and authors from different disciplines and countries for discussing with us and broadening our minds.

A special thank goes to the German Ministry of Education and Research and the German Aerospace Center (DLR) for the constructive and productive collaboration. This volume would not have been possible without the numerous contributors and their commitment (in alphabetical order): Hannah Bleher, Sophie Engelhardt, Tabea Ott, Max Tretter and Jason Valdez. We also thank Frank Schindler from Springer for the pleasant and successful cooperation and his excellent care.

References

Academy of Sciences Leopoldina. 2017. Ethical and Legal Assessment of Genome Editing in Research on Human Cells. https://www.leopoldina.org/uploads/tx_leopublication/2017_Diskussionspapier_GenomeEditing.pdf. Accessed 10 January 2017.

Academy of Sciences Leopoldina, et al. 2015. *The Opportunities and Limits of Genome Editing.* Halle: Academy of Sciences Leopoldina, et al.

Baltimore, David, et al. 2015. A Prudent Path forward for Genomic Engineering and Germline Gene Modification. *Science* 348: 36–38.

Berlin-Brandenburg Academy of Sciences and Humanities (BBAW). 2015. *Human Genome Surgery – towards a Responsible Evaluation of a New Technology: Analysis by the Interdisciplinary Research Group Gene Technology Report.* Berlin: Berlin-Brandenburg Academy of Sciences and Humanities.

Birnbacher, Dieter. 2014. *Naturalness. Is the "Natural" Preferable to the "Artificial"?* Blue Ridge Summit: UPA.

Braun, Matthias, and Peter Dabrock. 2016. "I bet you won't": The Science–Society Wager on Gene Editing Techniques. *EMBO Reports* 17: 279–280.

Braun, Matthias, and Peter Dabrock. 2017. Mind the Gaps! Towards an Ethical Framework for Genome Editing. *EMBO Reports* 19: 197–200.

Caplan, Arthur L., et al. 2015. No Time to Waste – the Ethical Challenges Created by CRISPR. *EMBO Reports* 16: 1421–1426.

Center for Genetics and Society. 2015. CRISPR/Cas Gene-Editing Technique Holds Great Promise, but Research Moratorium Makes Sense Pending Further Study. http://www.geneticsandsociety.org/article.php?id=8759. Accessed 31 December 2016.

Cyranowski, David, and Sara Reardon. 2015. Embryo Editing Sparks Epic Debate. *Nature* 520: 593–595.

Dabrock, Peter. 2009. Playing God? Synthetic Biology as a Theological and Ethical Challenge. *Systems and Synthetic Biology* 3: 47–54.

Dabrock, Peter, et al. 2013. A Primer to 'Bio-Objects': New Challenges at the Interface of Science, Technology and Society. *Systems and Synthetic Biology* 7: 1–6.

Dow, Lukas E. 2015. Modelling Disease In Vivo With CRISPR/Cas9. *Trends in Molecular Medicine* 21: 609–621.

Gasiunas, Giedrius, and Virginijus Siksnys. 2013. RNA-Dependent DNA Endonuclease Cas9 of the CRISPR System: Holy Grail of Genome Editing? *Trends in Microbiology* 21: 562–567.

German Ethics Council. 2017. Germline Intervention in the Human Embryo: German Ethics Council Calls for Global Political Debate and International Regulation. http://www.ethikrat.org/files/recommendation-germline-intervention-in-the-human-embryo.pdf. Accessed 30 September 2017.

Gross, Michael. 2015. Bacterial Scissors to Edit Human Embryos? *Current Biology* 25: R439–R447.

Gyngell, Christopher, et al. 2017. The Ethics of Germline Gene Editing. *Journal of Applied Philosophy*. doi:10.1111/japp.12249.

Hurlbut, J. Benjamin. 2015. Limits of Responsibility: Genome Editing, Asilomar, and the Politics of Deliberation. *Hastings Center Report*. doi:10.1002/hast.484.

Hyun, Insoo, and Catherine Osborn. 2017. Query the Merits of Embryo Editing for Reproductive Research Now. *Nature Biotechnology* 35: 1023–1025.

Ishii, Tetsuya. 2014. Potential Impact of Human Mitochondrial Replacement on Global Policy Regarding Germline Gene Modification. *Reproductive Biomedicine* 29: 150–155.

Joerden, Jan C., Eric Hilgendorf, and Felix Thiele. 2013. *Menschenwürde und Medizin. Ein interdisziplinäres Handbuch*. Berlin: Duncker & Humblot.

Lanphier, Edward. 2015. Don't Edit the Human Germ Line. *Nature* 519: 410–411.

Ledford, Heidi. 2015. CRISPR, the Disruptor. *Nature* 522: 20–42.

Liang, Puping, et al. 2015. CRISPR/Cas9-Mediated Gene Editing in Human Tripronuclear Zygotes. *Protein & Cell* 6: 363–372.

Liu, Zhen, et al. 2018. Cloning of Macaque Monkeys by Somatic Cell Nuclear Transfer. *Cell* 172: 881–887.

Ma, Hong, et al. 2017. Correction of a Pathogenic Gene Mutation in Human Embryos. *Nature* 548: 413–419.

Musunuru, Kiran. 2017. Genome Editing: The Recent History and Perspective in Cardiovascular Diseases. *Journal of the American College of Cardiology* 70: 2808–2821.

Nature Editorial. 2017. Humans 2.0. *Nature Biotechnology* 35: 993.

Nuffield Council on Bioethics. 2016. Genome Editing: An Ethical Review. http://nuffieldbioethics.org/wp-content/uploads/Genome-editing-an-ethical-review.pdf. Accessed 30 September 2017.

Ormond, Kelly E., et al. 2017. Human Germline Genome Editing. *The American Journal of Human Genetics* 101: 167–176.

Peters, Ted. 2014. *Playing God? Genetic Determinism and Human Freedom*. New York/London: Routlegde.

Rothhaar, Markus. 2015. *Die Menschenwürde als Prinzip des Rechts: Eine rechtsphilosophische Rekonstruktion*. Tübingen: Mohr Siebeck.

Schickl, Hannah, et al. 2014. Rechtsethische und rechtspolitische Herausforderungen im Umgang mit induzierten pluripotenten Stammzellen. *Medizinrecht* 32: 857–862.

Stier, Marco, and Bettina Schoene-Seifert. 2013. The Argument from Potentiality in the Embryo Protection Debate: Finally "Depotentialized"? *American Journal of Bioethics* 13: 19–27.

The Academy of Medical Sciences, et al. 2015. Genome Editing in Human Cells – Initial Joint Statement. http://www.wellcome.ac.uk/About-us/Policy/Spotlight-issues/Genome-editing/WTP059704.htm. Accessed 30 September 2016.

The National Academies of Sciences, Engineering, and Medicine. 2017. *Human Genome Editing. Science, Ethics and Governance*. Washington, DC: The National Academies Press.

I
Mapping the Current Legal Framing

Genome Editing and the Law
Some Remarks on Current Legal Challenges of CRISPR-Cas9

Stephan Rixen

Keywords

Biomedicine, Comparative Law, Constitution, Governance, Regulation

Abstract

The law is an essential parameter of the regulation of biomedicine. The patterns of law have to be reassessed in the light of new developments like CRISPR-Cas9. This is not about legal technicalities, rather at stake are fundamental evaluations of the pros and cons of CRISPR-Cas9. This is where morality and ethical judgment come into play. The task of the law is not to decide which moral view or ethical assessment is the true one. On the contrary, the law has to make moral plurality and ethical diversity operable within the constitutional framework of democracy. Looking at the legal situation in the UK may help to define an appropriate approach to regulate CRISPR-Cas9 in Germany.

1 The Function of Law in Biomedicine

"Between moral hazard and legal uncertainty. Ethical, legal and societal challenges of human gene editing". In fact, legal uncertainty and legal challenges cannot be identified without sidelong glances to ethics or societal effects. That is, they cannot be identified without appreciating the law's context. Although uncertainty is great and challenges are even greater, I assume that many people in academia, for example, philosophers, theologians or bioethicists, would prefer to lead the current debates

M. Braun et al. (Eds.), *Between Moral Hazard and Legal Uncertainty*, Technikzukünfte, Wissenschaft und Gesellschaft / Futures of Technology, Science and Society, https://doi.org/10.1007/978-3-658-22660-2_2

without lawyers. They may recall a famous word from Shakespeare's "Second Part of King Henry the Sixth" that to the present day strikingly illustrates just how popular lawyers are: Dick the Butcher says, "The first thing we do, let's kill all the lawyers" (Shakespeare 1993b, Act IV Scene 2).

Indeed, there seem to be many reasons for this attitude, even if, in most cases, they do not lead to the use of lethal weapons. One reason, amongst others, is a lawyer-specific way of coping with reality. According to Niklas Luhmann, the great sociologist whose training as a lawyer enabled him to assess the situation properly, throughout their education, lawyers learn to "stylise any issue to be decidable". If everything is decidable, this nevertheless does not mean everything has to be understood before we start with the making of rules and decisions. Lawyers – at least lawyers who reflect on their practice – are working under terms of preliminary definitiveness, because this is the characteristic of positive law: to give orientation as long as law-makers and regulators, especially parliaments, do not replace old legal provisions with new ones. Legal provisions are a way to structure or to change reality, and the views how this can be managed change with time, and so do legal provisions. This is not, as someone might see it from the viewpoint of cultural pessimism, "relativistic". It is the consequence of pluralistic democracies of the Western European tradition as oriented toward human rights. Pluralistic democracies prefer (or ought to prefer) piecemeal-thinking instead of a thinking in search of exclusive (and excluding) truth.

2 Legal Challenges

What then are the legal challenges of CRISPR-Cas9? "Clustered regularly interspaced short palindromic repeats" (CRISPR) and "CRISPR-associated protein 9" (Cas9) is a genetic engineering technology, metaphorically called "gene" or "molecular scissors". Cas9, a so-called endonuclease, is an enzyme able to cut DNA. Using CRISPR-Cas9, scientists can insert, delete or replace DNA in the genome of a living organism using engineered nucleases. "Science" hailed this as the 2015 Breakthrough of the Year (Science 2015).

Genome editing presents the opportunity to modify the genes of microorganisms, plants, animals, and most interesting for us, human beings, by so-called gene therapy or gene surgery. It is obvious that gene modification of plants and animals – according to the German Gene Technology Act genetically engineered organisms (GMOs) – may have, in the long run, an influence on agriculture, on food, and eventually, on the genome of human beings. Moreover, there seem to be a lot of opportunities for innovation in the field of reproductive medicine and "everyday" medicine, too. Medical experts say the new technology may help to develop therapies against viral infections, for example, HIV (that can lead to AIDS) or HIB (the Hepatitis B virus). According to experts, the new technology may help to eliminate genetic diseases which are, until today, incurable, for example, cystic fibrosis, haemophilia or the sickle-cell disease. Furthermore, in the field of oncological therapy, the new technology may help to improve the control of mutations and, thereby, may stop the growth of cancer cells. In brief, CRISPR-Cas9 seems to be a powerful multi-purpose medical weapon, and it may well be the beginning of many, many innovations which could eventually lead to evidence-based methods of medical therapy.

In other words, we are entering a land of hope and, near at hand, of glory (like Nobel Prizes in Medicine); but, at this stage, there is nothing fixed and fought out. There is a kind of gold-digger fever in biomedical research. However, until evidence-based therapies are established, we are moving in the field of basic research. Furthermore – and this is especially important from the lawyer's point of view – we are moving in the field of human experimentation and individual curative trials. The requirements of informed consent are high here, and, at least in (Western) Europe, control by state agencies is (more or less) strict. Many questions of medical malpractice law arise if these requirements are not considered. If we look at the typology of questions (for example, standards and limits of liability), these are well-known topics. New are the cases to which these standards and limits have to be applied.

However, for the time being, we have to deal with legal provisions that regulate the process of scientific research. Throughout Europe, we face a large number of norms that are very diverse in their details. In simple terms, they follow either a more prohibitive or a more liberal approach. Interventions on germ cells are especially a source of controversy, because of the consequences for future generations. Activists from the disability movement claim a "right to disability" as a version of a right to difference or a right not to be discriminated against. For them, CRISPR-Cas9 represents a smart kind of eugenics that endangers – not only in written law but in societal practice – the equality of people with or without special needs.

To conclude: If CRISPR-Cas9 leads to new methods of therapy, there will be many very concrete legal questions that need to and can be answered with the help of the established patterns of legal reasoning. Here, legal practice will be characterized by the evaluation of risks and benefits for the individual and the community, and by implementation of informed consent.

First of all, however, we have to know if law allows the initiation and performance of research on highly controversial genetic material, especially on human germ cells, and respectively, the human germline. Even research that crosses the border between species seems possible: as shown in a publication in "Cell", from January 2017, about "interspecies Chimerism" between humans and pigs, "based on CRISPR-Cas9-mediated zygote genome editing", as the authors put it (Wu et al. 2017). This could be one step toward healing, for example, type 1 diabetes, because this editing technology might rebuild a functioning pancreas that produces insulin.

3 Comparing Germany and the UK

3.1 UK

What is the legal situation regarding human germline research in the UK and Germany?

In a variation of a famous (or infamous) saying of former British Prime Minister Margaret Thatcher, one might say "There is no such thing as law." Law is a plurality of legal orders that are not completely determined, modified or replaced by international law (or supranational law, like EU law). There are about 200 states and – although the vanguard of the theory of international law does not agree – these states are still the basis of law enforcement. This is the situation in the field of health and biomedicine law, too. Let's have look at the UK.

In 1990, the Human Fertilisation and Embryology Act (as amended in 2008) installed a new regulator: the Human Fertilisation and Embryology Authority (HFEA). After the birth of Louise Brown in 1978, the world's first IVF (in vitro fertilisation) baby, discussions in the British public during the 1980s led to the creation of the HFEA. The HFEA is an executive, non-departmental public body of the Department of Health in the UK. "Non-departmental public body" (NDPB) is another name for quasi-autonomous non-governmental organisations (the so-called quangos). NDPBs are not an integral part of any government department. They carry out their work, as is often said, at arm's length from ministers, who are ultimately responsible to Parliament for the activities of bodies sponsored by their department. It is a form of organizational independence, a sort of specific distance between department and authority that restricts the possibilities of political influence.

The HFEA oversees all clinics in the United Kingdom that provide IVF, artificial insemination including the storage of human eggs, sperm or embryos (as defined by the Act) – and it also oversees human embryo research. The HFEA has to keep a database of every IVF treatment carried out, and a database relating to all cycles and uses of donated gametes (that is, eggs and sperm). The 2008 Amendment of the Human Fertilisation and Embryology Act especially ensured that all human embryos outside the body are subject to regulation. The Amendment, too, ensured regulation of "human-admixed" embryos created from a combination of human and animal genetic material for research. All in all, the HFEA is an authority that regulates assisted reproduction, and at the same time, one that controls scientific processes regardless of whether these follow from basic or applied research where embryos or gametes are concerned.

It was not a lawyer but a philosopher who made decisive steps towards the installation of the HFEA. In 1984, the "Report of the Committee of Inquiry into Human Fertilisation and Embryology", presided over by the Oxford philosopher Mary Warnock, recommended, in chapter 13 of the report, "the establishment of a new statutory licensing authority to regulate both research and [...] infertility services [...]." "We", the Commission says,

> "would therefore envisage a significant representation of scientific and medical interests among the membership. It would also need to have members experienced in the organization and provision of services. However, this is not exclusively, or even primarily, a medical or scientific body. It would also need to have members experienced in the organization and provision of services. It is concerned essentially with broader matters and with the protection of the public interest. If the public is to have confidence that this is an independent body, which is not to be unduly influenced by sectional interests, its membership must be wide-ranging and in particular the lay interests should be well represented. We recommend that there should be substantial lay representation on the statutory authority to regulate research and infertility services and that the chairman must be a lay person." (Warnock Report 1984, 76)

The Commission stated that there is "a wide diversity in moral feelings, whether these arise from religious, philosophical or humanist beliefs" (Warnock Report 1984, 2). The Commission continues: "The interests of those directly concerned, as well as those of society in general, demand that certain legal and ethical safeguards should be applied" (Warnock Report 1984, 75). The Commission does not accept strict utilitarian principles: "Moral questions, such as those with which we have been concerned are, by definition, questions that involve not only a calculation of consequences, but also strong sentiments with regard to the nature of the proposed activities themselves" (Warnock Report 1984, 2). The Commission, in other words, follows a model of moderate consequentialism which has to be dealt out in every case.

The Human Fertility and Embryology Act largely adopted the recommendations and installed a governing board – also called "the Authority" – and license committees. Lay representation is strong. It gives much room to the pluralistic views on biomedicine in every single act of decision-making. At present, the Authority is composed of ten members, six women and four men, some with migration backgrounds, all named by the Secretary of State for Health. The Chair is a finance and business professional with ties to the English branch of the National Health Service (NHS). Others come from the field journalism. There are women with positive or negative experiences of IVF. Four members have specific expertise in genetic research and medicine, and there is a Bishop of the Church of England. All persons have professional or personal experience in the field. The Anglican bishop, for example, was a member of the UK Gene Therapy Advisory Committee for nearly ten years.

The members of the Authority determine HFEA policies, review treatment and research licence application. The daily work is done in committees, especially the license or the appeal committees, which deal with non-routine complex matters, for example, human embryo research; here, the members of the Authority themselves make the decisions. Moreover, there is the HFEA's Scientific and Clinical Advances Advisory Committee (SCAAC), which reviews scientific and clinical developments affecting activities in which the HFEA has an interest, and provides recommendations about these developments. The Authority members are supported by a staff with, at present, about 65 members. Routine decisions are made by an Executive Licensing Panel composed of persons of the Authority's staff.

Decisions granting licenses, for example, regarding embryo research, are published, although the grounds provided are normally not very detailed. The Authority and the panels apply normative parameters that are consequentialist criteria of the middle range, and are focused on each specific case. For example, in February 2016, the HFEA licensed research done by the Francis Crick Institute in London. They were allowed to use CRISPR-Cas9 to modify the genes of developing embryos, with the goal of improving success rates and reducing miscarriages. It was the first time a national regulatory body has given the procedure the green light. The technology is not legal as such in the UK, but is only allowed for the scientists of that research institute.

The regulations and guidance of the HFEA "Code of Practice" try to focus on the necessity of research, only forbidding extreme activities like the implanting of non-human embryos in a woman. Thus, the HFEA emphasizes that research may only be carried out for one of the following purposes:

- increasing knowledge about serious disease or other serious conditions
- developing treatments for serious diseases or other serious medical conditions
- increasing knowledge about the causes of congenital diseases
- promoting the advancement of infertility treatment
- increasing knowledge about the causes of miscarriages
- developing more efficient techniques of contraception

- developing methods for detecting gene, chromosome or mitochondrion abnormalities in embryos before implantation
- increasing knowledge about the development of embryos.

In the end, the discussions in the Authority and the license committee follow the direction fixed by the Warnock Committee: "There must be some barriers that are not to be crossed, some limits fixed, [...]" (Warnock Report 1984, 2). "But in our pluralistic society it is not to be expected that any one set of principles can be enunciated to be completely accepted by everyone" (Warnock Report 1984, 2). The Committee accepts a "common moral position" avoiding extremes, and so does the HFEA.

This approach is accepted, for example, by the Nuffield Council on Bioethics' review on genome editing (Nuffield Council on Bioethics 2016). Without going into a detailed comparison here, the Nuffield Council on Bioethics may be thought of as the UK equivalent to the German Ethics Council. The Nuffield Council opposes "bioconservatism" and restricts the function of human rights "as grounds for claims against interference in the exercise of individual freedoms and, especially, against interference by public authorities" (Nuffield Council on Bioethics 2016, 28). Human rights, especially dignity, the Council says, are not a natural law-like "objective ground for making distinctions between acceptable and unacceptable uses of technology" (Nuffield Council on Bioethics 2016, 28), at least the Nuffield Council is sceptical in this regard. Accordingly, human rights in the biomedical field seem to be mere moral demands, not legal claims. The Council, then, emphasizes that "[d]emocratic governance purports to offer a procedurally legitimate solution to controversial questions in morally plural societies" (Nuffield Council on Bioethics 2016, 31).

The Council reminds us of the "strength of consequentialism" (Nuffield Council on Bioethics 2016, 29), because it focusses on benefits and harms. And, as the Council says, despite all differences in the ethical evaluation, "the content of moral judgements may show a reasonable degree of co-incidence" (Nuffield Council on Bioethics 2016, 31). This seems to be a variation of the Rawlsian "overlapping consensus" that is, first, moderately directed by consequentialism and that, second, is diffusely and rarely corrected by stronger moral positions in singular cases to come. This, to sum up, is the philosophical framework, approved by law, in which the HFEA makes its decisions.

3.2 Germany

What about the situation in Germany? First of all, we have to differentiate between the constitutional and the statutory level. The German "Act for Protection of Embryos", enacted in 1990, reads as follows in Section 5, paragraphs 1 and 2, on the "Artificial alteration of human germ line cells":

> 1. Anyone who artificially alters the genetic information of a human germ line cell will be punished with imprisonment up to five years or a fine.
> 2. Likewise anyone will be punished who uses a human germ cell with artificially altered genetic information for fertilisation.

Section 8, paragraph 3 of the Act gives a definition of germ line cells:

> Germ line cells, for the purpose of this Act, are all cells that, in one cell-line, lead from the fertilised egg cell to the egg and sperm cells of the resultant human being and, further, the egg cell from capture or penetration of the sperm cell until the ending of fertilisation by fusion of the nuclei.[1]

Although any modification is generally forbidden, there are some exemptions, as paragraph 4 shows (esp. no. 1 and no. 3):

> Paragraph 1 does not apply to
> 1. an artificial alteration of the genetic information of a germ cell situated outside the body, if any use of it for fertilisation has been ruled out,
> 2. an artificial alteration of the genetic information of a different body's germ line cell, that has been removed from a dead embryo, from a human being or from a deceased person, if it has ruled out that
>   a) they will be transferred to an embryo, fetus or human being or
>   b) a germ cell will originate from them, and likewise
> 3. inoculation, radiation, chemotherapeutic or other treatment by which an alteration of the genetic information of germ line cells is not intended.

The legislators tried to find a balance between a strong "pro-life" position, extending to very early forms of human life (that is always endowed with dignity), and the interests of scientific research and medical treatment. But, mainly, the scales are tipped against a wide concept of life and dignity. Life and dignity are inseparably linked. Human dignity is incessantly invoked when the question of life arises. This is problematic because with respect to the status quo of constitutional construction we cannot differentiate in an operable way between life and dignity.

1 Official translation slightly changed to express the meaning more equivalently.

But much more important is this: We are facing a conception of human rights where human rights are not only entitlements of born individuals but, at the same time, parts of an objective "order of values" (*Wertordnung*), as the German Federal Constitutional Court (1958) puts it. Construing human rights like this results in their becoming gateways for free-floating variations of moral or ethical debates without any compelling force to set limits. But, even worse, this is not a strong deontological position, since, in along consequentialist lines, the strong moral status of early forms of human life is called into question by scientific and medical interests. Thus, what confronts us, wrapped in the garb of legal language, is a veiled consequentialist approach with some deontological brakes – in my view: a symbolic rhetoric of deontology.

During the last years, legal scholars have developed different approaches, for example, a tiered or a gradual concept of dignity. But this is not compatible with the dominant reading of the guarantee of human dignity and human life in German legal scholarship. The Federal Constitutional Court has not yet ruled on the question, but there are strong voices in scholarship trying to adopt the Court's jurisprudence on abortion laws and extend them to biomedical situations. This is not unanimously accepted, but it has many adherents in society and amongst politicians and lawyers who deal with bioethics and biomedicine.

3.3 International Law

What about international law? As an example, let's have a look at the Council of Europe's 1997 "Convention for the Protection of Human Rights and Dignity of the Human Being with regard to the Application of Biology and Medicine: Convention on Human Rights and Biomedicine" (the so-called Oviedo-Convention). In Article 13, on "[i]nterventions on the human genome", we read: "An intervention seeking to modify the human genome may only be undertaken for preventive, diagnostic or therapeutic purposes and only if its aim is not to introduce any modification in the genome of any descendants."

Although the interpretation is controversial most voices deny that genome editing is allowed, according to this article, if it is associated with any intergenerational effect. Hence, an improvement of the human genome that is accompanied by a long-term effect beyond the individual concerned is not allowed. The ban is quite extensive, because it suffices that the intervention is seeking to modify the human genome. However, the broad margin of application is not relevant for the UK and Germany, because, for different reasons, they did not sign and ratify the Convention as did, for example, Denmark or Switzerland. The UK did not want to correct their science-friendly approach, whereas German policy-makers, who had critical attitude towards biomedicine, feared that the Act on the Protection of Embryos would have had to be amended in a "liberal", science-friendly way.

4 The Future of Legal Regulation

Is there any resort for finding a future regulatory way in Germany? Above all, I do not think international law will provide feasible solutions. International law, like the Council of Europe's Oviedo-Convention, is an obvious obstacle for biomedical and economic innovation. The UK law widely avoided putting hindrances in the way of scientific (and economic) progress and so did, for example, Chinese law, if there is anything in China resembling law as measured by western standards. For remember, it was a Chinese group who became the first to inject a person with cells containing genes edited using the revolutionary CRISPR-Cas9 technique. On October 28th, 2016, a team led by oncologist Lu You, at Sichuan University in Chengdu, delivered the modified cells into a patient with aggressive lung cancer as part of a clinical trial at the West China Hospital, also in Chengdu. Observers say this could spark a biomedical duel between China and the US. It is highly probable that the present administration in Washington will not support any international convention that is not also signed by China. We have to consider that one of the two corresponding research groups has its seat in California. The second research group works in Berlin, and I am convinced that, in the years to come, we will witness a re-interpretation, or an amendment, of the German Act on the Protection of Embryos. Furthermore, I am unwilling to believe that soft law arrangements will be an internationally effective instrument.

I think the only future model that will work is an HFEA-like piecemeal approach of reasoned, step by step muddling-through. However, the deontological unease is strong in Germany, and not only amongst politicians who deal with biomedicine. Bioethical debates seem to function as placeholders for deep-rooted, normative interpretations of our world. Too many lawyers in German academia, especially in the field of constitutional law, prefer a quasi-philosophical approach. They want to ground the grounds instead of doing their job: i.e., reflecting on how, in an operable way, the existing moral plurality can be turned to decisions. Their job is not to understand the grounds of the grounds of the grounds, but to invent feasible approaches of the middle range.

My opinion is that we should adopt the UK way of legal reasoning and deciding. What we need is a complementary mixture of consultative arenas and actors who are competent to make a decision. Consultative arenas have to give structure to the plurality of the discourse, and they must give specific advice. In Germany we have the German Ethics Council, a pluralistic consultative body that discusses, among other themes, the normative implications and consequences of genome editing (German Ethics Council 2017). I am convinced that this is not enough. We should transfer (not literally, but observing its institutional pattern) the HFEA structure of decision-making: A lay-dominated board that – on the basis of a new Act on Reproductive Medicine and Biomedical Research – formulates guidelines for panels that make decisions on research projects. The balancing of human rights (dignity, life, freedom of scientific research) would be delegated to plural panels, that would have to find solutions for every single case. This model of piecemeal decision-making, of incremental governance, is not a common approach in Germany, but we have the first beginnings of it, though hidden from great public attention.

One example is the so-called "GEKO", which is the German abbreviation for the "Committee of Gene Diagnostics". Another example is the "Central Ethics Committee on Stem Cell Research" (the German abbreviation is "ZES"). Unlike the German Ethics Council, which has only consultative functions, GEKO and ZES are decision-making bodies. But there is an important difference: ZES, and especially GEKO, are strongly dominated by experts of medicine and biology; there is no real lay participation in the sense of the Warnock Report. This overrepresentation of medical experts ought to be changed in order to make societal plurality, and the effects on society, institutionally visible. But, in principle, combining the models of GEKO and ZES with the HFEA model would result in a new body, which (like the HFEA) would have to be independent from strong political influence in its everyday work.

5 Conclusion

"The law hath not been dead, though it hath slept" (Shakespeare 1993a, Act II Scene 2). The situation of biomedical law is in fact much the same as described by Shakespeare in "Measure by Measure". German law has the opportunity to change via impulses from comparative law. It can thereby rise from the sleep of regulatory alternatives and come to regulatory awakening.

What Germany should do in the field of biomedical law can be learned by comparing it with the HFEA and its legal frame. If we accept that truth-driven approaches to biomedical problems are not compatible with the task of law in a pluralistic society, then we can imagine and implement a design of law which makes room for what is more important: the encounter of ethical diversity, and its transformation, with step by step regulation. This is the serving function of biomedical law: to make a dialogue of truths, views, and beliefs possible, in a mandatory way, by processing preliminary definitiveness in operable forms. This, I suspect, is not much in the way of "strong" doctrines of truth, but it is a lot in the way of a concept of biomedical law that has to avoid, due to plurality, any search for unique moral truth. If law succeeds in coordinating the struggle for bioethical truth in society, then we may see the law in its true colors.

Literature

Academy of Sciences Leopoldina, et al. 2015. Chancen und Grenzen des genome editing – The Opportunities and Limits of Genome Editing. https://www.leopoldina.org/en/publications/detailview/publication/chancen-und-grenzen-des-genome-editing-2015/. Accessed 27 April 2018.

Act for Protection of Embryos (The Embryo Protection Act). 1991. http://www.rki.de/SharedDocs/Gesetzestexte/Embryonenschutzgesetz_englisch.pdf?__blob=publicationFile. Accessed 27 April 2018.

Basic Law for the Federal Republic of Germany. 2018. Federal Ministry of Justice and Consumer Protection. https://www.gesetze-im-internet.de/englisch_gg/. Accessed 27 April 2018.

Benston, Shawna. 2016. CRISPR, a Crossroads in Genetic Intervention: Pitting the Right to Health against the Right to Disability. *Laws* 5: 1–15.

Braun, Kathrin. 2016. From Ethical Exceptionalism to Ethical Exceptions: The Rule and exception Model and the Changing Meaning of Ethics in German Bioregulation. *Bioethics* 17: 146–156.

Buchholz, Frank. 2017. Genom-Chirurgie. Zukünftige Medizin an der genetischen Wurzel des Übels. *Forschung & Lehre* 24: 26–27.

Cabinet Office. 2007. Public Bodies 2007. https://www.gov.uk/government/uploads/system/uploads/attachment_data/file/266224/PublicBodies2007.pdf. Accessed 27 April 2018.

Cabinet Office. 2016. Public Bodies 2016. https://www.gov.uk/government/publications/public-bodies-2016. Accessed 27 April 2018.

Callaway, Ewen. 2016. Embryo Editing Gets Green Light. UK Decision Sets Precedent for Research on Editing Genomes of Human Embryos. *Nature* 530: 18.

Carroll, Dana, and R. Alta Charo. 2015. The societal opportunities and challenges of genome editing. *Genome Biology* 16: 242.

Convention on Human Rights and Biomedicine. 1997. Convention for the Protection of Human Rights and Dignity of the Human Being with regard to the Application of Biology and Medicine. https://rm.coe.int/168007cf98. Accessed 27 April 2018.

Dederer, Hans-Georg. 2017. Mehr Fragen als Antworten – CRISPR-Cas9 aus rechtlicher Perspektive. *Forschung & Lehre* 24: 24–25.

Department of Health & Social Security. 1984. Report of the Committee of Inquiry into Human Fertilisation and Embryology (Warnock Report). http://hfeaarchive.uksouth.cloudapp.azure.com/www.hfea.gov.uk/docs/Warnock_Report_of_the_Committee_of_Inquiry_into_Human_Fertilisation_and_Embryology_1984.pdf. Accessed 27 April 2018.

Fateh-Moghadam, Bijan. 2017. Genome Editing als strafrechtliches Grundlagenproblem. *Zeitschrift für Medizinstrafrecht* 3: 146–156.

German Ethics Council. 2017. Germline Intervention in the Human Embryo: German Ethics Council Calls for Global Political Debate and International Regulation. http://www.ethikrat.org/publications/ad-hoc-recommendations/files/recommendation-germline-intervention-in-the-human-embryo.pdf. Accessed 27 April 2018.

German Federal Constitutional Court. 1958. Order of the First Senate of 15 January 1958. http://www.bverfg.de/e/rs19580115_1bvr040051.html. Accessed 27 April 2018.

Günther, Hans-Ludwig, et al. 2014. *Embryonenschutzgesetz. Kommentar.* Stuttgart: Kohlhammer.

HFEA Homepage. 2018. Human Fertilisation and Embryology Authority (HFEA). https://www.hfea.gov.uk/. Accessed 27 April 2018.

House of Commons (Public Administration Select Committee). 2014. Who's accountable? Relationships between Government and Arm's-length Bodies. https://publications.parliament.uk/pa/cm201415/cmselect/cmpubadm/110/110.pdf. Accessed 27 April 2018.

House of Lords Library. 2007. Library Note: Human Fertilisation and Embryology Bill (HL) [HL Bill 6, 2007–08]. https://researchbriefings.parliament.uk/ResearchBriefing/Summary/LLN-2007-007#fullreport. Accessed 27 April 2018.

Kersten, Jens. 2004. *Das Klonen von Menschen.* Tübingen: Mohr Siebeck.

Kersten, Jens. 2013. Technische und ethische Fragen der Medizin. In: *Besonderes Verwaltungsrecht, Vol. 2*, Ehlers, Dirk, et al. (eds.), § 54, Heidelberg: C.F. Müller.

Luhmann, Niklas. 1993. *Das Recht der Gesellschaft.* Frankfurt am Main: Suhrkamp.

NHS Homepage. 2018. National Health Service (NHS). https://www.nhs.uk/conditions/infertility/treatment/. Accessed 27 April 2018.

Nuffield Council on Bioethics. 2016. Genome Editing: An Ethical Review. http://nuffieldbioethics.org/wp-content/uploads/Genome-editing-an-ethical-review.pdf. Accessed 27 April 2018.

Porteus, Matthew H. 2015. Towards a New Era in Medicine: Therapeutic Genome Editing. *Genome Biology* 16: 286.

Rawls, John. 2005. *Political Liberalism. Expanded Edition.* New York: Columbia University Press.

Reich, Jens, et. al. 2015. Genomchirurgie beim Menschen – Zur verantwortlichen Bewertung einer neuen Technologie. Eine Analyse der interdisziplinären Arbeitsgruppe Gentechnologebericht bei der Berlin-Brandenburgischen Akademie der Wissenschaften. http://www.gentechnologiebericht.de/gen/publikationen/genomchirurgie-beim-menschen-2015. Accessed 27 April 2018.

Robertson, John A. 2004. Reproductive Technology in Germany and the United States: An Essay in Comparative Law and Bioethics. *Columbia Journal of Transnational Law* 43: 189–227.

Rolf, Sibylle. 2012. Human Embryos and Human Dignity: Differing Presuppositions in Human Embryo Research in Germany and Great Britain. *The Heythrop Journal* 53: 742–754.

Rosenau, Henning, et al. 2013. *Fortpflanzungsmedizingesetz Augsburg-Münchner-Entwurf (AME-FMedG).* Tübingen: Mohr Siebeck.

Science. 2015. And Science's 2015 Breakthrough of the Year Is… http://www.sciencemag.org/news/2015/12/and-science-s-2015-breakthrough-year. Accessed 19 February 2018.
Shakespeare, William. 1993a. Measure for Measure, Hylton, Jeremy (ed.), http://shakespeare.mit.edu/measure/index.html. Accessed 27 April 2018.
Shakespeare, William. 1993b. King Henry VI, Part 2, Hylton, Jeremy (ed.), http://shakespeare.mit.edu/2henryvi/index.html. Accessed 27 April 2018.
Shepperd, Sahar. 2016. The Revolutionary Gene Editing Technology: CRISPR-Cas9. *eHealth Law & Policy* 3: 11–13.
UK Donor Received Register Homepage. 2018. Donor Received Register. https://donorconceivedregister.org.uk/. Accessed 27 April 2018.
Wu, Jun, et al. 2017. Interspecies Chimerism with Mammalian Pluripotent Stem Cells. *Cell* 168: 473–486.

Integrating Ethical Standards into the Human Rights Framework

Considerations towards the Future Regulation of Genome Editing on an International Level[1]

Fruzsina Molnár-Gábor

Keywords

Binding Effect, International Law, International Regulation, Regulatory Instrument, UNESCO Declaration

Abstract

Various organisations have called for an international regulation of human genome editing, particularly of interventions into the human germline with the tool CRISPR-Cas9. In the last two years, several national and international organisations have themselves adopted standards that set the stage for what should and should not be allowed through this technology in an increasingly comprehensive manner. However, there is still no internationally harmonized approach to the application of genome editing in humans.

Based on the Constitution of UNESCO, which provides the organisation an ethical mandate, the present contribution aims at providing food for thought on the chances for and limits of a future UNESCO regulation on genome editing with particular regard to the integration of ethical standards into the human rights framework, the binding effect of a possible future instrument, the role of deliberation during its development, and selected long-term effects on the human rights system.

1 This Chapter is based on ideas generally developed in my doctoral thesis: Molnár-Gábor 2017. Part 2 particularly on pp. 100–103, Part 3.1. particularly on pp. 192–208, Part 4 particularly on pp. 289–312. Part 3.2. has also been analysed in the publication Molnár-Gábor 2014. For Part 4.1. two further publications are relevant: Molnár-Gábor 2015 and Molnár-Gábor 2016.

M. Braun et al. (Eds.), *Between Moral Hazard and Legal Uncertainty*, Technikzukünfte, Wissenschaft und Gesellschaft / Futures of Technology, Science and Society, https://doi.org/10.1007/978-3-658-22660-2_3

1 Introduction

1.1 Genome Editing: Scientific Achievements and Applications

Since the ground-breaking development of genome editing in 2011, biology has had an increasingly powerful molecular tool for rewriting genes and designing new genomes (Jinek et al. 2012). The special feature of the novel tool CRISPR-Cas9, the so-called "gene scissors", is that it can change genetic material easily, quickly, inexpensively and in a targeted manner. According to current scientific knowledge, this genome editing technology is so precise that interventions in the genome can generally no longer be recognized afterwards. In a way unlike any other technology, CRISPR has revolutionized large parts of the life sciences by privileging design over evolution (Esvelt 2014).

CRISPR is widely used in basic research and is increasingly gaining ground in many areas of applied sciences: i. e., not only in green biotechnology but also in the development of new tests in the pharmaceutical industry as well as in the context of gene therapy in humans (e. g. Yin et al. 2017; Gasperini et al. 2017; Bengtsson et al. 2017; Zhang et al. 2017). Its possible applications, encompassing changes in the human germline, or designing „selfish genes" (known as gene drives) to alter the genetic destiny of a species, all raise a wide range of ethical and social questions (Oye et al. 2014). In several countries, somatic gene therapy in humans has received socioethical and regulatory acceptance, because the genetic changes induced are not passed on to the next generation. It is precisely the technology's possible application to modify the human germline (either for therapeutic purposes or for purposes of enhancement of particular characteristics of an individual) that has raised the most severe ethico-moral concerns, initiating controversial debates about the need for its regulation (de Lecuona et al. 2017; Howard et al. 2017).

Firstly, ethico-moral considerations about genome editing are influenced by the technological limitations of its application. Generally, CRISPR-based gene therapy cannot yet provide a quick fix for the vast majority of diseases, which depend on many genes as well as environmental factors and lifestyles (UNESCO et al. 2015, 103). Secondly, safety is an unquestionable condition for its application to human beings where the intervention at issue is likely to have significant effects on the life of individuals (who could be considered to be "designed on demand" by someone else without their consent) and transmits their genome modifications to future generations. Eliminating a harmful predisposition might result in other problems, and impose other risks (on individuals and the human species itself) potentially as serious as the ones eliminated, due to unknown gene interactions and possible unintended consequences of modifying the human genome (UNESCO et al. 2015, 105). Thirdly, the destruction of embryos involved in some of these techniques revives the well-known controversy about the principle of respect for human life, and the related issue of the status of zygotes, embryos, and fetuses. Reaching a consensus on these issues seems practically impossible (UNESCO et al. 2015, Nr. 106; Academy of Sciences Leopoldina 2017). Finally, and closely connected to this third point: Compared to other genome editing technologies, editing the human germline using CRISPR might also impinge on the principle of respect for human dignity in several other ways. It weakens the idea that the differences among human beings, regardless of the measure of their endowment, are exactly what is presupposed and protected by the recognition of their equality. It could mean new forms of discrimination and stigmatization for those who cannot afford such enhancement, or simply do not want to resort to it. Furthermore, the arguments produced in favor of the so-called liberal eugenics do not trump the argument for taking account of the medical limits, also where CRISPR is concerned (UNESCO et al. 2015, Nr. 107–108).

Additionally, discussions about ethico-moral questions are hampered, in that the technology's application is increasingly driven forward by transnational initiatives as well as multi-centered research projects, and furthermore, in that the consequences of germline editing can have an effect on humankind beyond state borders. As research is being conducted worldwide and has global impact, debates on whether to prohibit or authorize systematic, transgenerational modifications to the human genome, and, if authorized in principle, the extent to which they would need to be limited by introducing certain conditions and restrictions, are to be increasingly discussed in a global manner. Thus, it is also a matter of debate whether and how the application of this technology can and should be regulated on an international level.[2] Fostering international regulation takes up the concept of the genome as the symbolic heritage of humanity, a phrase coined in Article 1 of UNESCO's Universal Declaration on the Human Genome and Human Rights from 1997 (UNESCO 1997), but that is of uncertain meaning in international law (Wolfrum 2009). However, with regard to international regulation, it has not yet been extensively discussed, which globally relevant institution should make the decision about the possibility of modifications to the genetic foundations of humankind. Furthermore, given ethico-moral, cultural plurality, it is foreseeable that regulatory questions about genome editing will be discussed in a highly controversial manner

2 At the Washington Summit, for instance, the prevailing attitude was that intervention in the human genome could not be regulated solely on the national level, but should also and primarily be regulated on an international level (Reardon 2015).

and that answers given will differ significantly. Furthermore, beyond international regulations, there is still room for discussion about whether there is an additional need for particular national regulations to respect specific historical experiences or cultural peculiarities. The German Ethics Council suggests, on the one hand, that local, regional and national groups, and professional audiences debate these questions. On the other hand, it proposes that attention also be drawn to these questions on a level in keeping with their global importance: the level of the politically organised world-community in the form of the United Nations (UN). The Council thus strongly promotes the adoption of global rules or conventions under international law by the UN (German Ethics Council 2017).

1.2 Aims of the Contribution

This contribution focuses on the United Nations Educational Scientific and Cultural Organisation's (UNESCO) work, with particular regard to the work of its International Bioethics Committee (IBC), as an international actor engaged in the regulation of genome editing de lege lata and possibly de lege ferenda. UNESCO's work on regulating genome editing so far will be presented, and some starting points for thoughts will be provided particularly on the possible future integration of ethico-moral standards into codified regulation when governing this technology on a comprehensive, transnational level. This may lead to a better understanding of how a harmonized transnational approach might be initiated concerning the application of this new technology, particularly in humans. The consideration of an increased binding effect of a regulatory instrument, achieved by its integration within itself of different normative standards, is a crucial first step towards the theoretical foundation of such a governance, and it should intersect with an engagement with the development of specific regulatory provisions for the application of CRISPR.

2 Background: The Relationship between Ethics and the Law in the Context of the International Regulation of Biotechnology

Ethics provides a reflective examination of practiced norms and values, having morals and the morality of actors (i. e., in real, lived attitudes to morality) as its object (Auer 2005; Vöneky 2010; Marckmann et al. 2012). In this definition, the term "moral" is used descriptively and refers to an empirically comprehensible normative belief system, which includes non-legal and non-conventional behavioral norms that are factually valid in a certain society, or social group, at a certain time (Vöneky 2010). The role of ethics is to provide arguments for the justification and validity, or for the evaluation of moral acts (Vöneky 2010; von der Pfordten 2011; Spranger 2010).

Law must firstly be distinguished from ethics as a system of reflection over the behavioral norms of moral, having a certain consistency and possessing the ability to establish new possibilities of action as well as to impose obligations on the addressees. (Hart 1994; Vöneky 2010). A legal system is effective if a part of those concerned accepts and follows the norms (von der Pfordten 2008). Based on effectivity as such modern positive law can fulfill its duty to stabilize expectations and to serve as a basis for trust in society (Luhmann 1999; Vöneky 2010).

The border between ethics and law can never be fluid, since moral views, as such, are never positive law, only condensing to legal provisions if they fulfill the formal and material conditions of the latter (Spranger 2010; ethics as a continuous control instance of law: Kirchhof 2006). Though both ethics and law answer the question of how to act, law refers to the positive law, the de facto legal order in force, whilst ethics provides help with moral decisions when answering this question (Kirchhof 2006). They do, however, interact with one another, and their generating and incorporating effect on each other is of the utmost relevance (Spranger 2010).

First, law can be generated from morals, or from ethically grounded "ought" norms, or it can open up to ethics, by means of which an ethical principle can be used to concretize an undefined legal term (Spranger 2010; von der Pfordten 2005). There is also the converse possibility that law influences morality, or morals and ethics. Second, the significance of moral values can motivate their legal incorporation, though the law's justification for demanding compliance with moral norms can by all means be questioned (von der Pfordten 2011). Once ethical principles are incorporated in law, the question arises about how far they are actually part of positive law: Is their incorporation immanent to law, or does it go further than law, being understood as the incorporation of an extra-legal valuation (von der Pfordten 2011)? If ethical principles are so deeply incorporated that they are to be considered a constituent part of law, then, as a result, there may be a certain relief from moral burdens for the law's addressees (von der Pfordten 2005). At the same time, it is doubtful how far such relief can be reached, owing to the difficulty of codifying ethical norms and the demanding work of proving compliance with them (Schöne-Seifert 2005).

In cases of controversial regulatory matters, a juridification of ethical standards is to be supported in the interest of legal certainty (Spranger 2010; von der Pfordten 2011). Ethical standards can contribute to increased objectivity and practical effect by connecting them to generally binding positive legal standards (Spranger 2010). With regard to the ethico-moral issues of genome editing, it therefore seems particularly justified that the law should be open to or based on ethico-moral judgments, while the degree, manner and scope of such processes would require additional analysis.

Scientific research that could have such fundamental effects on humanity's self-image, as genome editing might have, and discussions related to its regulation, must be thus embedded in society. It is not only an internal affair of the scientific community or the legislator. Nor is it only a matter for one country alone – not just because research happens in international networks but, foremost, because the consequences of these research activities affect everyone. This is the reason, as the German Ethics Council states, why the scientific community must endeavor to engage in open-ended discussions with all relevant groups of the international public. At the same time, global political institutions can and must find ways and initiate processes for discussing the numerous, as yet unanswered questions about, and the possible consequences of, systematic genome manipulations through genome editing. This has to be done in an intensive, differentiated, and above all, international manner, with the objective of drawing up the necessary regulatory standards (German Ethics Council 2017). Because of this, when integrating ethical standards on genome editing into regulatory initiatives, their understanding as a mirror of accepted social morals should also be taken into account (Jasanoff 2005; Tanner 2012). The various actors, and groups of actors, involved in genome editing or affected by its application, reflect particular narratives, which are influenced by individual societies' manners of coping with the novelty of science and technology, including various decision-making structures. On the other hand, the different groups of actors exhibit certain, specific traits of understandings of ethico-moral standards, that are only partly defined by their belonging to a certain society. Thus, there is an urgent mandate for the international community to both provide frameworks for the crystallization and deliberation of the relevant actors' ethico-moral standards, as well as structures for their transition into codified instruments of governance (all while taking into account their different societal embeddings and determinacies on various levels).

3 Genome Editing: A Snapshot of the International Regulatory Initiatives of UNESCO

Because the application and effects of genome editing technologies are becoming increasingly global, scientific and international organisations have established guidelines in order to develop harmonized approaches across borders, despite differences in national regulatory solutions. Actors in the area of public international law (such as the Council of Europe or UNESCO, in its function as a sub-organisation of the UN engaged, among others, in bioethical issues) have recently called for a temporary ban on genetic editing of the human germline, and have developed guidelines for the application of genome editing technology (DH-BIO 2015).

In this chapter I concentrate on the work of UNESCO, with its ethical mandate, based on its Constitution, and the influence of its work on possible future standard-setting activity concerning genome editing.

3.1 The Ethical Mandate of UNESCO

UNESCO perceives an ethical mandate in accordance with its statutory mission of promoting collaboration in science, as codified in Article 1 of UNESCO's Constitution (UNESCO 1946). It fulfils this statutory mission by discussing questions of bioethics and the ethics of science and technology, and by addressing ethical concerns about unbridled scientific development. On the other hand, it takes account of the concern that the rapid pace of scientific development has already outrun various dialogues going on in wider society about the impacts of such development (UNESCO 2008).

Within this mandate, the work of UNESCO includes four areas: It provides a forum for ethical debates, it acts as a standard-setter, and it promotes the building up of capacities as well as education and raising awareness (Bank et al. 2008, 8). The strategic situation, afforded by the multidisciplinary and multicultural setting of UNESCO as well as by UNESCO being the only UN organisation that has competences both in social and human sciences, allows UNESCO to function as a prominent catalyst of a global ethics framework that builds on universal principles, which allow all humanity to enjoy the benefits of science (UNESCO 2005). As a result of its standard-setting work, which is the most important facet of UNESCO's ethical mandate, three universal declarations on bioethical issues have been adopted in 1997, 2003 and in 2005.

Although these declarations represent international soft law, as such, they nevertheless have a considerable factual binding effect on states that adopt these documents (Neuhold 2005). Additionally, UNESCO has always been keen on integrating its standard-setting soft law work into the international human rights framework (Molnár-Gábor 2017). UNESCO explicitly acknowledges that, beyond rooting ethics in philosophical reflections, it has to be based on the framework of human rights (without considering human rights as a subject to ethics-based analysis). Grounding ethical means on the principle of human dignity aims at introducing law-driven approaches to bioethical issues in accordance with human rights, and attempts to introduce bioethics on the contractual basis of human rights (UNESCO 2008). The worldwide political consensus on the importance of protecting human rights might have a harmonising influence in dealing with (bio)ethical challenges of science and technology, once these are regarded as part of the universal human rights framework. This being said, the overarching aim of promoting collaboration among nations through science, as declared in the UNESCO Constitution, follows the ultimate goal of protecting human rights and fundamental freedoms (Molnár-Gábor 2017).

3.2 Standard-Setting Activities of UNESCO until now in the Field of Genome Editing

In response to rapid advancements in genetics and genomics, the IBC has decided to update its reflections on the human genome and human rights.[3] Beyond this, it also derives its responsibility from the fact that, in a symbolic sense, the human genome is "the heritage of humanity" (UNESCO 1997, Art. 1). In keeping with this definition, it is acknowledged that advancements of science and technology in this field entail a responsibility towards humankind, which has to be met not only by states and governments, but also by the international community as a whole. Perceiving this responsibility manifested in the "Report of the IBC on Updating its Reflection on the Human Genome and Human Rights" 2015 (UNESCO et al. 2015).

Five ethical principles and societal challenges were addressed in the report: respect for autonomy and privacy; justice, solidarity and non-discrimination; the understanding of illness and health; the cultural, social and economic context of science; and responsibility towards future generations (UNESCO et al. 2015, Art. 10–34). When applying these principles to practices, the precautionary principle should be respected as an overarching yardstick, ensuring that substantial consensus of the scientific community on the safety of new technological applications is the premise for any further consideration (UNESCO et al. 2015, Art. 117).

As it commonly does in its reports, the IBC made recommendations to various addressees. States and governments are called on, among other things, to become legislatively active on different levels, such as agreeing on a moratorium on genome

---

3 "The International Bioethics Committee of UNESCO should contribute to the dissemination of the principles set out in this Declaration and to the further examination of issues raised by their applications and by the evolution of the technologies in question" (UNESCO 1997, Art. 24).

engineering of the human germline, at least as long as the safety and efficacy of the procedures are not adequately proven as treatments. Even among scientists no agreement has been reached about the threshold of what could be considered as "acceptably safe and effective". At the same time, by acknowledging the existence of different perspectives and normative standards, the debate still remains open on the ethical acceptability of these techniques (UNESCO et al. 2015, Art. 118). Nevertheless, the IBC recognizes that scientific advancements in this field are likely to offer unprecedented tools for fighting diseases. Therefore, it is crucial to acknowledge that these opportunities should never become the privilege of the few, since the heritage of humanity entails sharing both in responsibilities and in benefits (UNESCO et al. 2015).

States and governments are also called upon to encourage, through the means of national legislation as well as international regulations, the adoption of rules especially regarding the modification of the human genome and the production and destruction of human embryos, and to adopt further legislative measures to ensure the appropriate application and governance of new genetic technologies on a general level. In addition, there would be good reason to revisit the existing UNESCO Declarations on bioethical issues and update some applications. However, according to the IBC, the cogency of codified principles should remain untouched (UNESCO et al. 2015).

The community of scientists and related regulatory bodies are called on to strengthen and participate in international fora, in order to update research and share information on the efficacy, safety, and consequences of new technologies related to the human genome; to set and share high quality standards for service delivery in genetics; and to renounce the pursuit of spectacular experiments that do not comply with the respect of fundamental human rights and universal ethical standards (UNESCO et al. 2015). Media and educators are called on to raise awareness and promote higher levels of health and scientific literacy, and among other things, to disseminate and strengthen the idea that scientific advancements in biomedicine entail responsibilities, to demarcate what should be accepted and allowed, which cannot solely be left to market forces of demand and supply (UNESCO et al. 2015). Last but not least, economic actors and for-profit companies are called on to comply with principles and regulations that ensure the highest standard of quality and safety for consumers, and to refrain from circumventing one country's restrictions, by taking advantage of weaker rules in other countries, in order to maximize profit (UNESCO et al. 2015).

Reports of the IBC on various bioethical issues have often set the stage for regulatory instruments of UNESCO in the past. Although the work programme of the IBC for 2018–2019 received a different focus, hitherto, all declarations of UNESCO on bioethical issues can be traced back to preparatory work provided by the IBC in the form of well-founded and detailed reports. Given the repeatedly articulated need for the international regulation of genome editing, particularly human germline editing, there is urgent reason to consider the potential of international standard-setting in this field, by a possible future UNESCO declaration. Such a declaration would be characterized by specific and due respect both for relevant ethical and legal standards, and would consider ways to ensure the integration of already existing standards, beyond those established by the IBC, as well as their responsible bodies during this work.

4 Considerations towards the Future International Governance of Genome Editing by UNESCO

4.1 Integrating Ethical Standards into the Human Rights Framework?

Firstly, the integration of ethical standards reflecting certain moral views into the legal framework of human rights has to be considered on a general level.

On a general level, all UNESCO declarations on bioethical topics have so far been incorporated into the framework of human dignity and human rights (Molnár-Gábor 2017). In their preambles and operative parts, these declarations refer to numerous international conventions and human rights treaties, inter alia to the Universal Declaration on Human Rights and both International Covenants (UN 1948; UN 1976a; UN 1976b). However, whether the integration of ethical standards into the human rights framework would be an adequate solution for regulating biomedicine is a matter of debate (Baker 1998; Thomasma 1997; Callahan 1996), requiring further thinking where the regulation of genome editing is specifically concerned.

In this respect, it should be taken into account that, as disciplines, ethics and law both refer to homonymic concepts and terms, a fact that does not necessarily mean they share an understanding of those notions, or have identical concepts. Criticisms mostly articulate that, together with a human rights language, a different methodology of solving conflicts in bioethics also becomes accepted. Firstly, a vertical viewpoint, typical for balancing human rights, might also be integrated into ethics (Sperling 2008). This can run counter to the traditionally horizontal viewpoint in ethical argumentation. Legal language too often implies the infringements of rights, while the language of ethics presupposes mutual dialog between actors, without assuming that rights were previously infringed upon, or that obligations have not been fulfilled. Secondly, ethical decisions can also be made between two legally acceptable solutions. Finally, and importantly, fulfilling legal standards is not always enough for making ethical decisions.

There are few areas, in which there is effectively a closer interface between law and ethics than in the area of medicine. The marked characteristic of the international law of biotechnology is the increasing and diverse consideration of ethics. Ethics has taken on a role, in this regulation, that is greater than its usual norm-initiating and norm-interpreting one. With particular consideration of the relationship of human rights and bioethics, UNESCO states: "Human rights law contains provisions that are analogous to the principles that flow from analysis of moral obligations implicit in doctor-patient relationships, which is the starting point, for example, of much of the Anglo-American bioethics literature, as well as the bioethics traditions in other communities" (UNESCO et al. 1994a, IV.1).[4]

In accordance with the rooting of the ethical mandate in the promotion of collaboration among nations through science, as declared in Article 1 of UNESCO's Constitution, the mandate's overarching aim is the protection of human rights and fundamental freedoms (Molnár-Gábor 2017).

However, it is questionable whether the human rights framework as such is an appropriate one for promoting a common understanding that is foundational for protecting the human being in the context of new technologies like genome editing. By integrating bioethical considerations into the human rights framework, UNESCO has often established new principles and standards in its already existing bioethical declarations. The content of the standards thus established is a mixture of statements of moral principles and statements in the form of human rights discourse. Ashcroft differentiates three broad classes of statements in the articles of bioethical declarations (Ashcroft 2008). Besides principle-based statements, and prescriptive statements similar to classical human rights proclamations, there are also "articulating norms", bridging the gap between principle-based statements and human rights norms (e.g. "Human dignity, human rights and fundamental freedoms are

4 Historically, there are many parallels in the development of medical ethics as the source of bioethics and human rights. From the very beginning the essential factor for the promotion of both fields has been the relief of human suffering. Most of all, in international humanitarian law, the norms of medical ethics overlap with the norms of international customary and contractual law, both standardizing the treatment of wounded civilians and prisoners in armed conflicts. Prominent personalities, who made considerable contributions to the shaping of human rights standards, were often medical doctors like John Locke (Faunce 2005, 173–174).

to be fully respected" [UNESCO 2005, Art. 3 (1)]). This sort of article underlines the difference between the declarations and other human rights instruments.

So far, however, the definition and weighting of these principles in the declarations of UNESCO has often been left open, seldom providing the necessary international governance of challenges in the life sciences. Beyond this lack of guidance for the weighting of the principles, critics also complain that the way bioethics is represented in the declarations (i. e., solely as a string of principles) is incomprehensible and of little help in special, challenging situations (Benatar 2005). Also, the declarations hardly formulate the *rights* of the persons concerned, but rather principles or prohibitions for their protection. If rights themselves are formulated in the declarations, it is in a vague and minimalistic way, making their application in new situations unclear (Benatar 2005).

4.2 The Binding Effect of a Future Instrument

Secondly, the integration of certain ethical standards into the human rights framework has to be regarded with view to the strengthening of the *de facto*, non-legal binding effect of a future possibly declaratory instrument, this means how far the instrument the praxis of its addressees actually influences.

A future international declaration on genome editing of UNESCO, which would qualify as a non-binding instrument, might appear appropriate, because it allows the broadest consent among member states. Furthermore, it suits the fast-changing context of medicine. Also, there is always a chance that declarations would later become binding, just like the UN Universal Declaration on Human Rights, which was adopted as a resolution of the General Assembly, was further specified by human rights treaties and now qualifies as customary international law (Charlesworth 2008, 14). In contrast, declarations and resolutions qualify first as soft law.

Soft law are abstract-general rules of conduct, that have been adopted by entities of international law but that cannot be assigned to any formal source of law and that are not legally binding in a direct manner (that is they do not comply with the formal-material conditions of legal bindingness) (Friedrich 2013; Molnár-Gábor 2012). However, soft law can achieve relevance according to its actual non-legal binding force, since it has an important practical binding effect on states that have endorsed such instruments. Because of this soft law cannot be reduced to qualify solely as recommendations of non-legal nature completely missing any legal effect. The importance of treaties as sources of international law remains uncontested. Their scope of application, however, is often not general enough for the governance of issues with prominent ethico-moral relevance. On the contrary, soft law promises a flexible, adequate and up-to-date standardization of bioethics and thus scores by a quality- and specificity-based approach (Boyle 1999). This can lead to a higher acceptance among the addressees. In addition, the usually transparent and increased deliberations by the development of soft law documents based on the active and dynamic involvement of various actors in the design process can increase the authority and through this the factual legitimacy of these instruments.

Contrary to legal binding force, which might be given or not, the *de facto* binding effect of such an instrument can have different levels (Molnár-Gábor 2017). One way to increase the binding effect is the integration of already existing international bioethical opinions and their established standards on genome editing into a future instrument. Standards in the field of genome editing have been lately developed foremost by actors of private law, international research institutes or within the framework of specific projects. The International Summit on Human Gene Editing unifying experts and advocacy groups released a position statement in 2015 in which they demand the resolution of ethical and safety issues before applying these technologies. Also, among others, the International Society for Stem Cell Research (ISSCR 2015), The Hinxton Group (2015), The National Academy of Sciences (2017), and the European Academies Science Advisory Council (2017) have released statements for the time being.

The integration of these already existing standards into a future international instrument could be achieved through opening clauses or preambles, each obtaining different prospects. As for the opening clauses, their application would only be possible under two conditions. First, the standards of the guidelines to be integrated must be obviously ethically justified. Second, these standards must not violate public international law and human rights (Vöneky 2012). However, opening clauses are usually formulated in a general manner and do not provide the possibility to differentiate between the instruments they integrate so that a broad integration of all guidelines without a previous approval would therefore might not be preferable.

The second way to integrate existing standards into a declaratory instrument would be based on the preambles of these instruments in which a reference to important and approved documents could be made. Preambles are part of the respective international instrument independent from whether this is a treaty or it is of declaratory nature because both the declaratory and the decisive part of an instrument construct together its regulatory context as defined in Article 31 of the Vienna Convention of the Law of the Treaties (Thürer 2009, 33). This of course does not imply an inherent normativity of the preamble, not even by international treaties, because the states cannot derive directly a mandate or an obligation from preambles.

Preambles however, contribute to the determination of the purpose of the instrument because they help to interpret the text and to capture the significance and meaning of it – and this again both by treaties or conventions of legally binding nature and by legally non-binding declarations (Gros Espiell 2009). As for capturing the significance of an instrument and defining its purpose, these might also include the integration of existing standards and guidelines that determine the genetic and genomic praxis on an international level.

Through the integration of such relevant international guidelines, a future possibly soft law instrument of declaratory nature from UNESCO would gain definitiveness itself and could fulfill its role better as a guide to the interpretation of international law itself and of national regulations (Thürer 2009). However, further research is needed on how far specific international guidelines actually represent ethico-moral and cultural standards.

4.3 The Practical Integration of Ethical Standards into a New Declaratory Instrument: The Role of Deliberation

According to UNESCO, genome editing affects the essence of human nature, because it dramatically changes the possibility of making interventions into human genetic material, and the human germline respectively, in ways that affect future generations and the human genetic code. Although this evaluation of the technology can itself be subject to debate, the integration of ethical standards (per se as well as reflected in already existing standards) into the above outlined human rights framework still has to be considered on a practical level.

By governing genome editing in the human rights framework, UNESCO could already make a statement towards the application of that status to humans, which is visible through the lenses of human rights. This already applies certain boundaries, concerning should and should not be allowed, in reflections on genome editing. Thinking of human rights not only as legal rights, but also as moral rights, and also as part of international law as a moral order, does reflect (as we all know) certain ethical standards. This way, only an ethical pluralism that resonates with human rights can be aspired to in the regulation of bioethical issues. This does not exclude ethical pluralism as such, but draws clear boundaries concerning how far such a pluralism may be acceptable against the backdrop of positive law. Furthermore, along these lines, the human rights framework could indeed function as a "minimum standard" of bioethics. In accordance with the understanding of ethics sketched above, in Part 2, this can only be legitimate if the development of any new regulation is based on wide public discussions and thorough deliberation.

The legitimacy of a declaration can be enhanced by increasing participation in the decision-making process, and by means of an efficient decision-making process. Beyond participation in IBC work groups, the Intergovernmental Bioethics Committee or other committees that prepare the biomedical standard-setting of UNESCO and the adoption of its declarations (including consultations within the UN), the organisation and procedure of consultations with actors outside the UN should also be approved.

If a call for the international control of genetic analyses arises, efforts should first be made to obtain the necessary, available knowledge. Rules and regulations have to be recorded precisely. In genome editing, for example, this could mean the exact mapping of the action sequences of various workflows. Once a detailed analysis is available, and the characteristics of a new technology have been fully grasped in terms of its challenges, a basis is provided that can give credibility to the discourse processes. The IBC can function as the first actor for the gathering of scientific information and knowledge, since this corresponds to its task assignment. Through a variety of approaches in the work of UNESCO, and in particular the IBC, there is regular clarity in the UNESCO's efforts to capture, in its declarations, the most recent state of human genetics research, for the achievement of effective governance. Making a contribution to cooperation in science is to be realized in a forward-looking manner, with UNESCO assuming the role of an international coordinator.[5] Although the reports of the IBC in the field of genetics are particularly sensitive to the normative challenges of genetic analysis, the statute allows the involvement of external experts in development processes, which the IBC has so far not widely used.

A more effective involvement of international players in genome medicine in the decision-making process would be appropriate for meeting the demand for topicality in current work on genetics. Although the IBC has had, for most of the time (during the development of its declarations until now), two permanent members, who were also involved in the Human Genome Organisation, other interest groups in genome research could also be better represented in the future (Friedrich 2013). This is particularly important with regard to the legitimacy of decisions, since these actors can later also become addressees of an adopted instrument. The role of UNESCO, and in particular, of the IBC could then be better focused on providing

5 "[...], the need for UNESCO to play a role in the international debate on the ethics of genetics continues. Indeed, given the rapid developments in human genetics with increasing numbers of genetic data banks being established, controversial uses of genetic data, and increasing non-medical use, it could be argued that this role is now more important than ever and one which UNESCO is uniquely placed to fulfil" (UNESCO et al. 2002, I.6).

orientational knowledge (Mittelstrass 2002; Fassbender 2006, 14) and integrating deliberative processes into the human rights frame of reference, in order to maintain and steer this framework for discourse and deliberation.

4.4 Effects on the Human Rights System

In order to develop international regulation of human genome editing, the chances and limits of UNESCO's methodology, for integrating ethical standards into the framework of human rights, has also to be considered with regard to its effects on the human rights system.

For this reason, when it comes to the future regulation of genome editing, further questions should be addressed such as whether a concretization of human rights, through the integration of bioethical standards into their framework, establishes or should establish a new generation of human rights (UNESCO et al. 1994b, 59; biomedical human rights, or even international subjective rights: Peters 2013), or whether this specification could only result in the further development of human rights for a specific context.

On the one hand, apart from states, key players in the application of genome editing technologies (foremost individuals such as medical doctors and researchers, but also the persons affected by this technology) could be handled as addressees of rights and obligations in the regulation of the application of this technology. The specification of human rights (in relation to the patients' collective, or in relation to concerned persons as a vulnerable group), and the increasing involvement of individual actors and institutions within the human rights framework, may encourage an understanding of the developed, human-rights-based principles, or human rights themselves, increasingly developed as inter-individual rights. On the other hand, this understanding could be supported by the integration of ethico-moral standards of medical ethics into the human rights framework by the regulation of genome editing, since ethical standards also primarily consider relationships between individuals or groups.

Integrating ethical standards into the human rights framework might thus lead to the further development of human rights as international subjective rights, possibly leading in consequence to an "empirical turn" (Borry et al. 2015) in human rights *de lege ferenda*.

However, the extent to which biotechnology will demand to identify the legal personality of the individual through a dynamic set of rights also in international law (Peters 2013), and thus to structure the protection of persons affected (in the face of rapid technological development) to some extent independently of willingness on the part of the established actors of international law (i. e., the states [Grzeszick 2005]), will also be a question of human genome editing's own technological development.

References

Academy of Sciences Leopoldina. 2017. Ethical and Legal Assessment of Genome Editing in Research on Human Cells. http://www.leopoldina.org/uploads/tx_leopublication/2017_Diskussionspapier_GenomeEditing.pdf. Accessed 27 February 2018.

Ashcroft, Richard. 2008. The Troubled Relationship between Bioethics and Human Rights. In *Law and Bioethics. Current Legal Issues. Volume 11*, Freeman, Michael (ed.), 31–51. Oxford: Oxford University Press.

Auer, Karl Heinz. 2005. *Das Menschenbild als rechtsethische Dimension der Jurisprudenz*. Wien: LIT.

Baker, Robert. 1998. A Theory of International Bioethics: The Negotiable and the Non-Negotiable. *Kennedy Institute of Ethics Journal* 8: 233–273

Bank, Roland, and Friederike Foltz. 2010. United Nations Educational, Scientific and Cultural Organisation (UNESCO). Max Planck Encyclopedia of Public International Law. http://opil.ouplaw.com/view/10.1093/law:epil/9780199231690/law-9780199231690-e553?rskey=-jmZEyW&result=1&prd=EPIL. Accessed 27 February 2018.

Benatar, David. 2005. The Trouble with Universal Declarations. *Developing World Bioethics* 5: 220–224.

Bengtsson, Niclas, et al. 2017. Corrigendum: Muscle-Specific CRISPR/Cas9 Dystrophin Gene Editing Ameliorates Pathophysiology in a Mouse Model for Duchenne Muscular Dystrophy. *Nature Communications*. Doi: 10.1038/ncomms16007.

Borry, Pascal, et al. 2015. The Birth of the Empirical Turn in Bioethics. *Bioethics* 19: 49–71

Boyle, Alan. 1999. Some Reflections on the Relationship of Treaties and Soft Law. *International and Comparative Law Quarterly* 48: 901–913.

Callahan, Daniel. 1996. Escaping from Legalism. Is it Possible? *The Hastings Center Report* 26: 34–35.

Charlesworth, Hilary. 2008. Universal Declaration of Human Rights (1948). Max Planck Encyclopedia of Public International Law. http://opil.ouplaw.com/view/10.1093/law:epil/9780199231690/law-9780199231690-e887. Accessed 27 February 2018.

Committee on Bioethics (DH-BIO). 2015. Statement on Genome Editing Technologies. https://rm.coe.int/168049034a. Accessed 27 February 2018.

de Lecuona, Itziar, et al. 2017. Gene Editing in Humans: Towards a Global and Inclusive Debate for Responsible Research. *Yale Journal of Biology and Medicine* 90: 673–681.

Esvelt, Kevin M. 2014. Design over Evolution. http://www.sculptingevolution.org/cas9/design-over-evolution. Accessed 24 February 2018.

European Academies Science Advisory Council. 2017. Genome editing: scientific opportunities, public interests and policy options in the European Union. https://www.easac.eu/fileadmin/PDF_s/reports_statements/Genome_Editing/EASAC_Report_31_on_Genome_Editing.pdf. Accessed 30 March 2018.

Fassbender, Bardo. 2006. Wissen als Grundlage staatlichen Handels. In *Handbuch des Staatsrechts der Bundesrepublik Deutschland*, Vol. IV: Aufgaben des Staates. 3rd ed., Isensee, Joseph, and Paul Kirchhof (eds.), 243–312.Heidelberg: C. F. Müller.

Faunce, Thomas A. 2005. Will International Human Rights Subsume Medical Ethics? Intersections in the UNESCO Universal Bioethics Declaration. *Journal of Medical Ethics* 3: 173–178.

Friedrich, Jürgen. 2013. *International Environmental "Soft Law"*. Heidelberg/Berlin: Springer.

Gasperini, Molly, et al. 2017. CRISPR/Cas9-Mediated Scanning for Regulatory Elements Required for HPRT1 Expression via Thousands of Large, Programmed Genomic Deletions. *American Journal of Human Genetics* 101: 192–205.

German Ethics Council. 2017. Germline Intervention in the Human Embryo: German Ethics Council Calls for Global Political Debate and International Regulation. Ad Hoc Recommendation. http://www.ethikrat.org/publications/ad-hoc-recommendations/files/recommendation-germline-intervention-in-the-human-embryo.pdf. Accessed 28 February 2018.

Gros Espiell, Héctor. 2009. The Preamble. In *The UNESCO Universal Declaration on Bioethics and Human Rights. Background, Principles and Application*, ten Have, Henk, and Michele Jean (eds.), 57–66. Paris: UNESCO.

Grzeszick, Bernd. 2005. Rechte des Einzelnen im Völkerrecht. *Archiv des Völkerrechts* 43: 312–344.

Hart, Herbert L. A. 1994. *The Concept of Law.* 2nd ed. Oxford: Oxford University Press.

Howard, Heidi C., et al. 2018. One Small Edit for Humans, One Giant Edit for Humankind? Points and Questions to Consider for a Responsible Way forward for Gene Editing in Humans. *European Journal of Human Genetics* 26: 1–11.

International Society for Stem Cell Research (ISSCR). 2015. The ISSCR Statement on Human Germline Genome Editing. http://www.isscr.org/professional-resources/news-publicationsss/isscr-news-articles/article-listing/2015/03/19/statement-on-human-germline-genome-modification. Accessed 28 February 2018.

Jasanoff, Sheila. 2005. *Designs on Nature. Science and Democracy in Europe and the United States.* Princeton: Princeton University Press.

Jinek, Martin, et al. 2012. A Programmable Dual-RNA-Guided DNA Endonuclease in Adaptive Bacterial Immunity. *Science* 337: 816–821.

Oye, Kenneth A., et al. 2014. Regulating Gene Drives. *Science* 345: 626–628

Kirchhof, Paul. 2006. Medizin zwischen Ethik, Recht und Vorbehalt des Möglichen. In *Humaniora. Medizin* – Recht – Geschichte. Festschrift für Adolf Laufs zum 70. Geburtstag, Kern, Bernd-Rüdiger, et al. (eds.), 931–952. Berlin/Heidelberg/New York: Springer.

Luhmann, Niklas. 1999. *Ausdifferenzierung des Rechts. Beiträge zur Rechtssoziologie und Rechtstheorie.* Frankfurt am Main: Suhrkamp.

Marckmann, Georg. 2012. Allgemeine Einführung in die medizinische Ethik. In *Ethik in der Medizin: Ein Studienbuch.* 4th ed., Wiesing, Urban (ed.), 23–37. Stuttgart: Reclam.

Mittelstrass, Jürgen. 1989. Glanz und Elend der Geisteswissenschaften. http://oops.uni-oldenburg.de/1192/1/ur27.pdf. Accessed 27 February 2018.

Molnár-Gábor, Fruzsina. 2017. *Die internationale Steuerung der Biotechnologie am Beispiel des Umgangs mit neuen genetischen Analysen.* Berlin: Duncker & Humblot.

Molnár-Gábor, Fruzsina. 2016. Bioethics. Max Planck Encyclopedia of Comparative Constitutional Law. http://oxcon.ouplaw.com/view/10.1093/law:mpeccol/law-mpeccol-e78?rskey=mDrxZx&result=7&prd=MPECCOL. Accessed 27 February 2018.

Molnár-Gábor, Fruzsina. 2015. Bioethics. Max Planck Encyclopedia of Public International Law. http://opil.ouplaw.com/view/10.1093/law:epil/9780199231690/law-9780199231690-e2150?prd=EPIL. Accessed 27 February 2018.

Molnár-Gábor, Fruzsina. 2014. The Ethical Mandate of UNESCO. In *Max Planck Yearbook of United Nations Law*, Lachenmann, Frauke, et al. (eds.), 332–367. Leiden/Boston: Brill.

Molnár-Gábor, Fruzsina. 2012. Die Herausforderung der medizinischen Entwicklung für das internationale soft law am Beispiel der Totalsequenzierung des menschlichen Genoms. *Zeitschrift für ausländisches öffentliches Recht und Völkerrecht* 72: 695–737.

Neuhold, Hanspeter. 2005. The Inadequacy of Law-Making by International Treaties: "Soft law" as an Alternative? In *Developments of International Law in Treaty-Making*, Wolfrum, Rüdiger, and Volker Röben (eds.), 39–52. Heidelberg/Berlin: Springer.

Peters, Anne. 2013. *Jenseits der Menschenrechte*. Tübingen: Mohr Siebeck.

Reardon, Sara. 2015. Global Summit Reveals Divergent Views on Human Gene Editing. Representatives Discuss Ethical, Social and Legal Issues that Unite and Divide Them. https://www.nature.com/news/global-summit-reveals-divergent-views-on-human-gene-editing-1.18971. Accessed 27 February 2018.

Schöne-Seifert, Bettina. 2005. Medizinethik. In *Angewandte Ethik. Die Bereichsethiken und ihre theoretische Fundierung.* 2nd ed., Nida-Rümelin, Julian (ed.), 690–803. Stuttgart: Alfred Körner.

Sperling, Daniel. 2008. A Rights-Based Relationship and Its Troubling Implications. In *Law and Bioethics. Current Legal Issues. Volume 11*, Freeman, Michael (ed.), 52–78. Oxford: Oxford University Press.

Spranger, Tade M. 2010. *Recht und Bioethik*. Tübingen: Mohr Siebeck.

Tanner, Klaus. 2011. Bioethik als kulturelle Praxis. In *Marsilius-Kolleg 2010/2011*, Kräusslich, Hans-Georg, and Wolfgang Schluchter (eds.), 213–223. Heidelberg: Winter.

The Hinxton Group. 2015. Statement on Genome Editing Technologies and Human Germline Genetic Modification. http://www.hinxtongroup.org/hinxton2015_statement.pdf. Accessed 27 February 2018.

The National Academies of Sciences, Engineering, and Medicine. 2017. *Human Genome Editing. Science, Ethics and Governance*. Washington, DC: The National Academies Press.

Thomasma, David C. 1997. Bioethics and International Human Rights. *The Journal of Law, Medicine & Ethics* 25: 295–306.

Thürer, Daniel. 2009. Soft Law. Max Planck Encyclopedia of Public International Law. http://opil.ouplaw.com/view/10.1093/law:epil/9780199231690/law-9780199231690-e1469?prd=EPIL. Accessed 27 February 2018.

United Nations (UN). 1976a. International Covenant on Civil and Political Rights, G.A. res. 2200A (XXI). http://hrlibrary.umn.edu/instree/b3ccpr.htm. Accessed 27 February 2018.

United Nations (UN). 1976b. International Covenant on Economic, Social and Cultural Rights. G.A. res. 2200A (XXI). http://hrlibrary.umn.edu/instree/b3ccpr.htm. Accessed 27 February 2018.

United Nations (UN). 1948. Universal Declaration of Human Rights. http://www.un-documents.net/a3r217a.htm. Accessed 27 February 2018.

United Nations Education, Scientific and Cultural Organisation (UNESCO). 2008. Ethics of Science and Technology at UNESCO. http://unesdoc.unesco.org/images/0016/001600/160021e.pdf. Accessed 27 February 2018.

United Nations Education, Scientific and Cultural Organisation (UNESCO). 2005. Universal Declaration on Bioethics and Human Rights. http://portal.unesco.org/en/ev.php-

URL_ID=31058&URL_DO=DO_TOPIC&URL_SECTION=201.html. Accessed 27 February 2018.
United Nations Education, Scientific and Cultural Organisation (UNESCO). 2003. Universal Declaration on Human Genetic Data. http://portal.unesco.org/en/ev.php-URL_ID=17720&URL_DO=DO_TOPIC&URL_SECTION=201.html. Accessed 27 February 2018.
United Nations Education, Scientific and Cultural Organisation (UNESCO). 1997. Universal Declaration on the Human Genome and Human Rights. http://portal.unesco.org/en/ev.php-URL_ID=13177&URL_DO=DO_TOPIC&URL_SECTION=201.html. Accessed 27 February 2018.
United Nations Education, Scientific and Cultural Organisation (UNESCO). 1946. Constitution of the United Nations Educational, Scientific and Cultural Organisation. http://portal.unesco.org/en/ev.php-URL_ID=15244&URL_DO=DO_TOPIC&URL_SECTION=201.html. Accessed 27 February 2018.
United Nations Education, Scientific and Cultural Organisation (UNESCO); International Bioethics Comittee (IBC). 2015. Report of the IBC on Updating Its Reflections on the Human Genome and Human Rights. http://unesdoc.unesco.org/images/0023/002332/233258E.pdf. Accessed 27 February 2018.
United Nations Education, Scientific and Cultural Organisation (UNESCO), and International Bioethics Comittee (IBC). 2002. Human Genetic Data: Preliminary Study by the ICB on Its Collection, Procession, Storage and Use. http://www.eubios.info/UNESCO/ibc2002.pdf. Accessed 27 February 2018.
United Nations Education, Scientific and Cultural Organisation (UNESCO), and International Bioethics Comittee (IBC). 1994a. Report on Human Gene Therapy. http://unesdoc.unesco.org/images/0013/001323/132347e.pdf. Accessed 27 February 2018.
United Nations Education, Scientific and Cultural Organisation (UNESCO), and International Bioethics Comittee (IBC). 1994b. Proceedings 1994. http://unesdoc.unesco.org/images/0012/001203/120377mo.pdf. Accessed 27 February 2018.
von der Pfordten, Dietmar. 2011. *Rechtsethik*. 2nd ed. München: C. H. Beck.
von der Pfordten, Dietmar. 2008. Was ist Recht? Eine philosophische Perspektive. In *Rechtsphilosophie im 21. Jahrhundert*, Brugger, Winfried, et al. (eds.). 261–285. Frankfurt am Main: Suhrkamp.
von der Pfordten, Dietmar. 2005. Rechtsethik. In *Angewandte Ethik. Die Bereichsethiken und ihre theoretische Fundierung*. 2nd ed., Nida-Rümelin, Julian (ed.), 202–301. Stuttgart: Alfred Körner.
Yin Chaoran. 2017. In Vivo Excision of HIV-1 Provirus by saCas9 and Multiplex Single-Guided RNAs in Animal Models. *Molecular Therapy* 25: 1168–1186.
Vöneky, Silja. 2010. *Recht, Moral und Ethik*. Tübingen: Mohr Siebeck.
Vöneky, Silja. (2012). Ethische Standards im Wissenschaftsrecht. In *Wissenschaft und Ethik. Wissenschaftsrecht. Beiheft 21*, Löwer, Wolfgang, and Klaus F. Gärditz (eds.), 68–96. Tübingen: Mohr Siebeck.
Wolfrum, Rüdiger. 2009. Common Heritage of Mankind. Max Planck Encyclopedia of Public International Law. http://opil.ouplaw.com/view/10.1093/law:epil/9780199231690/law-9780199231690-e1149. Accessed 27 February 2018.
Zhang, Yongping. 2017. Gene Editing in T Cell Therapy. *Journal of Genetics and Genomics* 44: 415–422.

II
Theoretical Challenges

Prospects of Human Germline Modification by CRISPR-Cas9 – an Ethicist's View

Dieter Birnbacher

Keywords

Embryo, Ethics, Individualisation, Nature, Normative Boundaries

Abstract

Genome editing holds the promise of revolutionizing many fields in which human interventions have hitherto proved to be insufficient to meet major global challenges, like nutrition and environmental protection. However, it is controversial how far this method might also be applied to the human germline with a view to preventing the transmission of serious genetic diseases to offspring. While there is a near-consensus that genome editing, at the present stage of science, should not be applied clinically, it is unclear whether this also extends to clinically oriented research. It is argued that among the arguments against interventions in the human germline there is only one that is sufficiently strong to be practically relevant: The argument that it is doubtful whether the challenge of off-target effects with potentially fatal health consequences can be met. Since nearly all objectives of human germline genome editing can be attained by PGD (except for the rare case of homozygous parents with respect to the relevant genes), there are good grounds for directing clinically oriented research toward the improvement of this alternative method, which has been proven to be without substantial risks and which involves no ethical problems over and above those involved in genome editing.

M. Braun et al. (Eds.), *Between Moral Hazard and Legal Uncertainty*, Technikzukünfte, Wissenschaft und Gesellschaft / Futures of Technology, Science and Society, https://doi.org/10.1007/978-3-658-22660-2_4

1 Introduction: The Limited Half-time of Normative Boundaries in Bioethics

Bioethical discussions are often confronted with the criticism that they systematically come too late because the progress of science is so swift that it outdistances ethical reflection and presents it with unalterable faits accomplis. Certainly, there is some truth in this criticism. Nevertheless, there are good examples of a more proactive role of bioethical debate. A paradigmatic example is the debate on reproductive cloning, which anticipated the reality years in advance. One of the first German contributions to the problems of cloning humans appeared as early as 1979 (Kliemt 1979). There was even legislation on human cloning long before the prospect of cloning mammals became a reality. However, this may be a special case because of the presence of cloning in science fiction. The more frequent situation is that technical progress takes bioethics by surprise, as, for example, in the case of "Social Freezing", the prospect of which had not particularly interested medical ethics before its arrival.

It is not only the agenda of bioethics that is frequently changed by technical advances, but also the dominant bioethical valuations and attitudes. New developments relativize seemingly well-established positions, or even reverse them, because they were not thought of or could not be thought of beforehand. Relevant examples are the modification of the dominant bioethical verdict on interventions in the human germline by the advent of mitochondrial replacement therapy (MRT), the transplantation of the nucleus (pronuclear transfer, PNT) or the chromosomes (maternal spindle transfer, MST) of a woman's oocyte with a genetic risk into the enucleated oocyte of another woman. In this way, a small part of the child's DNA derives from a female donor and not from its parents, i. e. exactly that part that bears the genetic risk. This method was still unknown at the time when interventions in the human germ-line were declared illegal in many countries. In retrospect, and in light of scientific progress, this judgment appears to be too crude. In Great Britain, at least, clinical applications of the method were officially approved in 2015, against the backdrop of a corresponding vote from the relevant commissions.

Was this the "crossing of the Rubicon" where germline modification is concerned? The European Group on Ethics (the bioethical advisory commission of the European Union) used just these words, in its comments on the breakthrough, with reference to the new method's future possibilities of germline interventions, making special reference to the most efficient among them, CRISPR-Cas9 (European Group on Ethics 2015). The question arises how far the approval of MRT prejudges an analogous approval of the application of CRISPR-Cas9 to the human germline, especially in light of the successful experiments by Liang et al. (2015) on non-viable human embryos and the more recent and still debated experiments by Ma et al. on viable embryos (Ma et al. 2017). There seems to be little time left for providing an answer.

There are obvious analogies between MRT and CRISPR-Cas9. Both are able to change the make-up of the human genome in an inheritable form. The primary objective in both cases is to prevent the transmission of serious genetic disease from parents to offspring. For both the question arises whether the benefits for parents and children are proportionate to the risks, including the risk of later generations inheriting the respective alterations.

The main dissimilarity between the methods is that MRT involves no direct intervention in the genome. With both methods, whole lumps of genetic information are transferred without intervening in their composition. Furthermore, the methods by which MRT is achieved constitute an established routine in reproductive medicine, whereas genetic intervention in the germline is still an experimental method. This is a significant ethical difference. In this respect, the existing analogies do not allow any substantial conclusions as to the ethical prospects of germline genome editing.

1.1 A Moratorium on Clinical Application but not on Clinically Oriented Research?

The existing state of opinion is characterized by a near-consensus that, at present, no attempts should be undertaken to apply the methods of human germline genome editing in a clinical context.

For the time being, the risks seem too high or too incalculable to justify applications that aim at correcting genetic anomalies and preventing the transmission of genetic diseases. Representative of this assessment is the position statement of the Summit on Human Gene Editing, issued in Washington D. C. in autumn 2015, in which the group of experts call for the resolution of ethical and safety issues before applying the new technologies. At the same time, the question arises whether a corresponding moratorium should hold for research directed at future clinical applications of germline modification by genome editing, including research directed at testing its safety.

This question, which seems much more controversial, can be split up into two questions: 1. Should research directed at genome editing of the human germline be legally permitted? 2. Should research directed at genome editing of the human germline be funded or otherwise supported by society?

It is easier to find an answer for the first question than for the second. Prohibiting research of a certain kind is always at odds with the principle of freedom of research, as contained in most constitutions of democratic states. Restrictions can only be justified by compelling reasons on the basis of high-ranking constitutional principles. It may be doubted, for example, whether the restrictions on research with human embryonic stem-cells, contained in the German Stem Cell Act, are compatible with the freedom of research stated in the German Constitution (cf. National Ethics Council 2007, 39–40). As far as I can see, there are no compelling reasons in sight where germline gene correction is concerned. Given the general guarantee of freedom of research, the threshold for research bans is, and should be, high. An opinion in this direction is also expressed in a relevant statement of the American Hinxton Group:

> "In the case of human genome editing, as with all science, it is important to target restrictive policy specifically to those dimensions of the research or its application that have proved to be unacceptable, and that these policies be proportionate to the magnitude of what is morally at stake." (Hinxton Group 2015)

Even the German Embryo Protection Act, which is not known to be particularly sympathetic to embryo research, shies away from prohibiting all forms of research on genome editing. According to sec. 5 (4) of the Act, "artificially altering the genetic information of a human germline cell" (sperm, egg cells, and their direct precursor cell stages) is perfectly legal, provided the respective cell is not used for reproductive purposes. While the Act prohibits all forms of experimenting on human embryos, it explicitly does not rule out genetic modification of gametes, pronuclei, or their precursors, if these are not transferred into a human embryo, fetus, or living human body. In light of the important role this Act accords to the potentiality principle, one may even speculate that even the research by Liang et al. (2015), with non-viable human embryos, might possibly be legal in Germany.

Another reason to be cautious in legally banning human germline genome editing is the difficulty of drawing, in the words of the European Group on Ethics, a "clear-cut distinction between basic and translational research" (European Group on Ethics 2015). In other words: A ban on research for clinical applications of germline genome editing might collaterally hamper possibly promising research with regard to fields that are not in the same way subject to ethical reservations – as for example basic research on early human development and human somatic gene therapy.

The second question deserves a more elaborate answer. Whether germline genome editing should be funded or otherwise supported by society essentially depends on the ethical chances and risks of the instruments eventually emerging from the research. "Chances" and "risks" should be understood here in an inclusive sense, i. e. without prejudging whether these are based on consequentialist considerations (such as human health and wellbeing), non-consequentialist considerations (such as human dignity or justice), or both. I will first consider the non-consequentialist ones.

2 Non-Consequentialist Arguments against Human Germline Genome Editing

The non-consequentialist arguments relevant to human germline genome editing are not new. Nearly all of them played a prominent role in the 1990s in relation to human cloning and to what then was called human germline therapy.

One of the most frequent, though at the same time most vulnerable, arguments is the argument from *human dignity*. Human dignity, so it is said, is compromised by "instrumentalizing" the individual by manipulating its genetic make-up. This is implausible for more or less obvious reasons. First of all, "instrumentalizing" an individual means that an individual is made the instrument of purposes other than its own, or apart from its own good. However, at least the standard application of genome editing is for the respective individual's own good. It is not the case that the individual or its good is "sacrificed" for the good of others.

This is true in spite of the fact that germline genome editing might have significant collateral consequences for others, especially for the respective individual's offspring, which might inherit the traits altered by genome editing. Nevertheless, it would be grossly misleading to say that the first individual in the chain of generations is made a means for the good of its descendants. If there is reason to assume that the intervention will be beneficial to later generations, it will also be beneficial to the first link in the chain.

More suitable candidates for the "instrumentalization" argument are the human embryos that would be used in research for germline genome editing. In the case of these embryos, at least two of the conditions are fulfilled that make an act of "instrumentalization" appear morally problematic: Namely, that first an individual is made an instrument for purposes other than its own, and second that doing this is in some way detrimental to the good of this individual, for example, by harming it or by reducing its chances for life or for a good life. The first condition is no doubt fulfilled in the case of embryo research that aims at developing technologies for human germline genome editing. Yet, as such, it does not justify the categorical verdict implied by the judgment that it violates human dignity. The damage done by "sacrificing" an embryo – regardless of whether it is left over from an in-vitro-fertilization or is specifically generated for research – must be balanced against the good expected from the development of gene modifications by developing an efficient instrument to prevent genetically transferred diseases. The second condition applies only if the termination of the life of a human embryo is considered to be a reduction of its good or its chances to have a good life. This seems problematic for conceptual reasons. A human embryo is not a person. It is a potential person. By terminating its life, it is prevented from existing as a human subject to whom the judgment, that it has a good or less than good life, can be meaningfully applied.

It is often thought that "instrumentalizing" a human being is, as such, a violation of human dignity. This is an oversimplification. The relationships between "instrumentalization" and violations of human dignity are more complex. This was already recognized by Kant, who was careful to make a distinction between treating a human individual as a means to an end and treating it "merely" as a means to an end. "Instrumentalization" can be a violation of human dignity only if one or several conditions, in addition to making someone a means to an end for others, are fulfilled (cf. Birnbacher 2008):

1. The "instrumentalized" individual suffers severe damage by whomever "instrumentalizes" it, either physically or mentally or both (as in torture, humiliation, and in severe cases of hostage-taking);
2. The "instrumentalized" individual is deprived of all, or nearly all, of its rights (as in slavery, or in the practice of buying and selling mercenaries across nations);
3. The instrumentalization involves the crippling of its specifically human capacities, such as the capacity for freedom, for rational thought, and for moral reasoning (as by torture, brain washing, or even prolonged imprisonment).

Nothing of the sort is involved in germline genome editing or in the relevant research settings. The conclusion seems inevitable. The argument of human dignity fails to be relevant in this context.

A second non-consequentialist argument is that the *natural* composition of the human genome should be treated as sacrosanct. By intervening, for preventive reasons, in the natural composition of the genome, germline genome editing in humans violates this principle.

However, the question is whether the underlying principle is at all plausible. It is true, germline genome editing intervenes in the genetic composition of the individual. Yet so do other accepted forms of genetic intervention, such as somatic gene therapy. Are the differences between the ways somatic and germline genome editing tinker with the natural course of events morally significant?

It was strongly asserted that there are important differences in one of the first reactions of German bioethicists to the emergence of the cloned sheep, "Dolly", in the 1980s:

> „Offensichtlich steht die freie Entfaltung der individuellen Person mit der Wahrung der Struktur der natürlichen Reproduktion in einem so engen Zusammenhang, dass um der Würde und Freiheit der einzelnen Person willen auch die Würde der mit der

> menschlichen Gattung verbundenen natürlichen Reproduktion respektiert werden muss."[1] (Eser et al. 1989, 236)

This statement brings out the full force of the underlying normative principle of naturalness: According to this position, by violating the "naturalness" of reproduction, germline genome editing would violate not only the dignity, but even the freedom of the individual.

However, it must be asked in what way naturalness can serve as a guarantee of dignity and freedom, especially in cases where letting nature take its course means condemning the individual to prolonged suffering and dependence. Obviously, this position is based on an unduly idealized image of the natural. This idealizing tendency is in conformity with the "bonus" that the concept of the natural receives in popular thinking (cf. Birnbacher 2014, 21). For example, natural risks are commonly less dreaded than anthropogenic risks. But this bonus more often than not seems undeserved. Natural risks such as diseases and other natural health risks seem to have a much greater impact on human life than anthropogenic risks such as criminal assaults.

In our case, however, naturalness is not a good argument anyway. Although, in everyday thinking, the natural generally receives a "bonus" over the artificial, this "bonus" is not conceded in all contexts with equal strength. In some contexts, the cultural and the civilized is clearly preferred to the natural, as in education and moral cultivation, in the enhancement of physical capacities and outer appearance by "natural" means (such as training and discipline) and, most relevantly, in medical therapy and prevention. In these contexts, the common devaluation of the artificial over the natural is, as it were, suspended.

Against this background, it can be said that genome editing for medical reasons, understood in a wide sense of the term, does not even infringe on popular conceptions of the natural. Artificiality in medical matters is far from being a taboo. Medical means may be as artificial as they like, if only they help.

Another problem of any attempt to base the moral doubtfulness of genome genome editing on a normative concept of naturalness is that interventions in the genome of species other than the human species only rarely meet with moral reservations. (A possible exception is the concept of *Würde der Kreatur* [dignity of living beings] introduced into the Swiss constitution on the occasion of increased use of genetically modified organisms in biomedical research.) Is there a morally

1 "Obviously, the unrestrained development of an individual person is so deeply connected with the protection of the structure of natural reproduction, that consequently for the sake of the dignity and freedom of the single person the dignity of the natural reproduction of the human species is to be respected."

significant difference between genetic interventions in humans and in other animals? In both cases, humans "play God" and alter the course of evolution. If it is *hubris* to deliberately alter the composition of the human genome, the practice of breeding animals to serve human needs and preferences should be illegitimate as well. However, this conclusion is accepted by only few of those who think that human germline genome editing violates the sacrosanctity of nature.

3 Consequentialist Arguments against Human Germline Genome Editing

Unlike non-consequentialist arguments, consequentialist arguments have an empirical basis, either in empirical fact or in empirical speculation. This is a strength and a weakness at the same time. Their strength is that their basis is publicly accessible, so that no secret knowledge or special intuition is necessary to recognize their relevance. Their weakness is that empirical assessments, when they address prospects for the future, often are hazardous, especially in dynamic fields such as biotechnology, and that in consequence, any consensus on the basis of consequentialist principles is as unlikely as a consensus on the basis of non-consequentialist ones.

It is interesting to note that more than one of the non-consequentialist arguments against germline genome editing mentioned have, as it were, a consequentialist analogue. In these analogues, it is not the objective state of affairs – the fact that a human germline is subjected to correction by genome editing – that is held to be critical, but rather their respective psychological reflections. These analogues refer to the responses, as it were, to this state of affairs in human minds, and especially in the minds of persons who either have been subjected to genome editing, when they were early embryos, or whose genome has been built by germline cells that have been subjected to genome editing before.

The first non-consequentialist argument we considered was the argument that genome editing is "instrumentalization". There is not only one but two psychological analogues of this argument. One is the argument that the individual resulting from germline genome editing might see itself as the victim, as it were, of the purposeful alteration of his or her "true nature". This may well be the case whenever the individual suffers from one or the other burden that he or she attributes to the intervention in his or her own original genetic constitution. Since the intervention was initiated by others, and with their respective purposes in mind, the individual may feel that the intervention was an illicit intrusion into his or her autonomy, similar to other kinds of interventions, such as the surgical fixation of sex in intersex children, or the irreversible inclusion into a certain religious denomination by circumcision. The individual might feel that it has fallen victim to manipulations by parents, doctors, and communities, who, with whatever good intentions, have made him or her a means to the fulfilment of their purposes, ideals, or belief systems.

I do not believe that this argument gets us very far. Experience shows that individuals generated even by the most exotic reproductive technologies rarely think of their origins, and only rarely feel disturbed by the knowledge that their characteristics are, to a substantial degree, the results of other people's strategic decisions. There is evidence that they develop no less "normally" than others (cf. Golombok et al. 1996; Colpin et al. 2002).

A second consequentialist argument, again psychological, is that the intrusion of strategic and technical rationality into the domain of reproduction might be felt, in Habermasian terms, to be an unwanted "colonization of the life-world". The intrusion of technology and the logic of means-end rationality into procreation, and in its wake, the corresponding extension of reproductive responsibility, might be seen as a disturbance of one of the last resorts of spontaneity in life. Perhaps it is this kind of consideration that made the European Parliament go so far, in 1982, as to postulate a "right to chance", with a correlative prohibition of interventions in the germline. Everyone should have the assurance that nobody has interfered with his or her genome without his or her explicit, informed consent.

Again, it is doubtful whether this consideration carries much weight. After all, a predetermination of physical characteristics is widely accepted if it is oriented towards the health and wellbeing of the individual child or the wider community, even if it is irreversible, as in the case of fetal surgery, prenatal therapy, or inoculation. With the possible exception of the fixation of sex in early childhood, these practices are usually retrospectively accepted by later individuals, especially where the health gains they achieve are evident. In the majority of cases, the individual will not develop a preference for not having been the "victim" of genetic correction, given that the genome editing process was directed at the correction of a serious anomaly and not intended as an expression, and imposition, of the parents', doctors', or the community's aesthetic or ideological tastes.

The second non-consequentialist argument, the argument that germline genome editing infringes on the inviolability of nature, has, again, a consequentialist analogue: The argument that interferences with the natural course of events reduce the range of options of the later individual, or, in Joel Feinberg's phrase, its "right to an open future". In fact, the individual may, under unfortunate circumstances, have feelings of regret over the changes it has undergone in the earliest stages of its development. It might have the thought that it would have fared better if it had been left alone and had developed without genetic interference.

However, if practiced to prevent serious genetically caused diseases (and not for purposes of enhancement), germline genome editing will extend the individual's future range of options instead of reducing it. As a rule, the individual can be expected to feel as little regret over these interventions as it will feel regret over all other therapies with irreversible consequences it has undergone in childhood, if only these have been undertaken in its own best interest.

This holds true, I think, in spite of the fact that, in psychological respects, interventions in the germline may have a special status. According to the theory of genetic exceptionalism at least, interventions in the genetic make-up of an individual, especially if they are irreversible, tend to be felt to affect the individual's essence to a higher degree than interventions in other physical characteristics. If this thesis is true, the correction of a genetic irregularity may be felt to be a more serious intrusion into the individual's identity than other interventions. Nevertheless, it is not to be expected that anyone will long for his or her "true" identity, defined by his or her "natural" genetic make-up, if this means a lifetime of suffering from a serious and incapacitating disease.

Another consequentialist argument, sometimes made against the ethical legitimacy of research on germline genome editing, concerns the potential social disruptiveness involved in research on human embryos. The background assumption is, that in most parts of the world, research on human embryos is morally controversial and is likely to remain so in the future. On the other hand, human germline genome editing will only be acceptable if it is tested and found to be sufficiently reliable. This requires a great deal of human embryo research, which, owing to the fairly fundamental differences in their developmental mechanisms, cannot be replaced by research on animal embryos (such as those of nonhuman primates).

Again, I personally doubt whether this consequentialist argument has sufficient force to question the legitimacy of research on germline genome editing in humans. First, embryo research could be conducted where it is already widely accepted, as it is by now in a substantial portion of the industrialized world. Second, embryo research is no longer as controversial as it was 20 years ago, when social and political opinion was influenced by religious doctrines to a greater extent than it is now. And third, the extent to which embryo research is accepted essentially depends on the plausibility of the ends for which it is undertaken. This plausibility is greatest where the resultant technology promises substantial gains in human health.

4 The Argument of Comparative Risk

There is, however, one consequentialist argument, that, in my view, calls into question the legitimacy of research directed at a potential clinical use of human germline genome editing: Namely, the improbability of meeting the challenge of making the method safe enough for clinical application. This improbability may be contested, especially with reference to examples from the history of technology. Quite a number of challenges, that were at first held to be insuperable, later proved to be sufficiently manageable to introduce the technology into general use. Nevertheless, it seems to me that the safety issue confronts human germline genome editing with a substantial, and not only temporary, problem. My impression is that it is grave enough to make research directed at clinical application of germline genome editing in humans seem problematic from the start.

I concede that this view may seem extreme. It certainly is contrary to what leading scientists in the field appear to believe (cf. Doudna et al. 2017) as well as to the perspective of The National Academies (2017), which, by formulating ethical conditions for its legitimacy, shows that it is, in principle, in favor of scientific research on potential clinical uses of human genome editing.

My view is motivated by two kinds of considerations:

1. The first aspect is the inherent risk profile of germline genome editing. The specific risk profile of germline genome editing is characterized by three general features, each of which exacerbates the risks of potential failures to attain the desired aim of successfully altering the composition of a human genome. (A) In contrast to somatic genome editing, correction of anomalies has to be attempted in the very early stages of a human individual's existence, at the stage of gametes, their precursors, the pronucleus, or the embryo in its very first stages of development. This implies that every single one of the individual's somatic cells will be altered by the intervention. A potential failure of the operation would leave traces in, and have potential impact on, the functioning of every single one of the individual's cells. (B) Furthermore, the alterations produced by genome editing are, as far as we can tell, irreversible. That implies that potential mistakes are irreparable. (C) And finally, alterations of the composition of the genome can be inherited by offspring, including a long series of later generations.
2. A second important aspect is the comparison of the risks of germline genome editing to those of the established alternatives for the prevention of genetic disease, especially PGD.

The risks of germline genome editing in humans compare unfavorably with those of PGD. Genetic analysis of the genome, and the consequent elimination of gametes or embryos with deleterious genomes, has proved to be manageable without risks to the resulting individual. What is more, germline genome editing itself involves PGD in screening out embryos for successful correction. Additional risks are introduced into the process by the genome editing process. So far, the undesired effects of this crucial step have not proved to be controllable. Until now, it has not been possible to introduce genetic alterations that are completely specific and do not also affect other parts of the genome. These "off-target effects" cannot be completely identified by PGD directed at edited genes, partly because of mosaic effects. Furthermore, genome editing interventions may fail to successfully target both alleles of a deleterious gene with the consequence that only one of the alleles is modified (cf. Araki et al. 2014).

On comparison, the question seems legitimate, why germline genome editing should be developed, given that, with PGD, there is a successful and nearly risk-free method available for preventing the transmission of serious genetic diseases. In my view, there are indeed good reasons to promote PGD rather than germline genome editing. A sober calculation of benefits and risks suggests that germline genome editing does not seem to be able to compete with PGD in its clinical prospects.

5 Objections

It might be objected that PGD, in its turn, has serious limitations, both in ethical and in medical respects.

The ethical objection is that PGD compares unfavorably with germline genome editing because it involves the "discarding" of human embryos whereas germline genome editing amounts to a purely therapeutic intervention: The embryos are "cured" instead of being subjected to a selection process, during which many of them are eliminated. Admittedly, the opposition between "elimination" and "therapy" as well as between the negative connotations of the first and the positive connotations of the latter is popular. Nevertheless, I do not think one should be much impressed by it.

First, it must be kept in mind that germline genome editing, too, involves the elimination of embryos, namely, when it becomes clear, in the screening process, that it is necessary to eliminate embryos in which genome editing has had significant off-target effects. Geneome editing has, in this respect, no ethical advantage over PGD.

Second, elimination of embryos in the course of a PGD procedure, given a medical indication, is widely accepted. It is in conformity with the legislation in most industrialized countries as well as in conformity with predominant moral views.

Most legislations presuppose a gradualist conception of the moral status of the human embryo, for example by grading criminal sanctions for abortion in proportion to the developmental status of the embryo or fetus. This is at variance with the officially absolutist doctrines of certain religious denominations, such as the Roman Catholic Church. Criminal sanctions for offences against embryonic life and integrity are generally far less grave than criminal sanctions for offences against adult human life and integrity. In some nations (as in Germany), abortion of the embryo in the first 14 days of development is exempted from legal consequences altogether.

Independently, and as a reconstruction of prevailing ethical views on embryo protection, a gradualist conception based on feelings of piety (comparable to the feelings towards deceased humans) seems, on the whole, intuitively more adequate than an absolutist conception (cf. Birnbacher 2011). A piety-based conception allows a gradation of protection in correspondence to the embryo's respective stage of development. The more the embryo reaches viability, the worthier of protection it is commonly held to be and the more it is seen as "one of us". By contrast, the principle of potentiality, which is in the background of the German Embryo Protection Act, implies an absolute right to protection from the start. Potentiality is a non-gradable property and applies to the blastocyst in the same way as to the viable fetus.

Another objection to the thesis that PGD may serve as an equivalent of germline genome editing is PGD's limited scope. This limitation is due to the fact that selection is unable to introduce anything new. It can only work with the genetic material provided by the genomes of parents.

This fact might prove to be a serious restriction with parents who both have the relevant, genetically caused disease and are homozygous with respect to the relevant genes. In this case, the parents have a 100 percent risk of having children who inherit their disease, regardless of whether the genetic anomaly is dominant or recessive. Not only Chorea Huntington, a late-onset disease with high penetrance caused by a dominant monogenetic factor, would resist the application of PGD, but also a much more frequent recessive anomaly like Cystic Fibrosis. In both cases, only germline genome editing could enable the couple to have a genetically related child without passing on the disease. The same holds for parents of whom one is homozygous with respect to a dominant genetic anomaly, or to parents with a genetic anomaly that severely limits fertility.

One may speculate whether or not the group of parents with this unfortunate genetic constellation, however small, will exert pressure to develop germline genome editing for clinical application in these extraordinary cases. To me, it is far from clear whether or not one should yield to this possible pressure. In a field of research with so many unknowns as genome editing, a risk-averse strategy, and a more reluctant approach to research in this direction, seems, all in all, ethically more adequate.

References

Araki, Motoko, and Tetsuya Ishii. 2014. International Regulatory Landscape and Integration of Corrective Genome Editing Into in Vitro Fertilization. *Reproductive Biology and Endocrinology* 12: 108.

Birnbacher, Dieter. 2008. Annäherungen an das Instrumentalisierungsverbot. In *Menschenwürde. Begründung, Konturen, Geschichte*, Brudermüller, Gerd, and Kurt Seelmann (eds.), 1–24. Würzburg: Königshausen & Neumann.

Birnbacher, Dieter. 2011. The Role of Complicity in the Ethics of Embryonic Stem Cell Research. In *Embryonic Stem Cells – Recent Advances in Pluripotent Stem Cell-Based Regenerative Medicine*, Atwood, Craig (ed.), 3–20. Zagreb: InTechnopen.

Birnbacher, Dieter. 2014. *Naturalness. Is the "Natural" Preferable to the "Artificial"?* Blue Ridge Summit: UPA.

Colpin, Hilde, and Stijn Soenen. 2002. Parenting and Psychosocial Development of IVF Children: a Follow-Up Study. *Human Reproduction* 17: 1116–1123.

National Ethics Council. 2007. Should the Stem Cell Act be Amended? Opinion. http://www.ethikrat.org/files/Opinion_Should_the_Stem_Cell_Law_be_amended.pdf. Accessed 15 February 2018.

Doudna, Jennifer, and Samuel Sternberg. 2017. *A Crack in Creation. The New Power to Control Evolution*. London: The Bodley Head.

Eser, Albin, et al. 1998. Klonierung beim Menschen. Biologische Grundlagen und ethische Bewertung. In *Hello Dolly? Über das Klonen*, Ach, Johann (ed.), 223–247. Frankfurt am Main: Suhrkamp.

European Group on Ethics in Science and New Technologies. 2015. Statement on Gene Editing. https://ec.europa.eu/research/ege/pdf/gene_editing_ege_statement.pdf. Accessed 15 February 2018.

Golombok, Susan, et al. 1996. Children: The European Study of Assisted Reproduction Families: Family Functioning and Child Development. *Human Reproduction* 11: 2324–2331.

Hinxton Group. 2015. Statement on Genome Editing Technologies and Human Germline Genetic Modification. http://www.hinxtongroup.org/hinxton2015_statement.pdf. Accessed 15 February 2018.

Kliemt, Hartmut. 1979. Normative Probleme der künstlichen Geschlechtsbestimmung und des „Klonens". *Zeitschrift für Rechtspolitik* 12: 165–169.

Liang, Puping ,et al. 2015. CRISPR/Cas9-Mediated Gene Editing in Human Tripronuclear Zygotes. *Protein & Cell* 6: 363–372.

Ma, Hong, et al. 2017. Correction of a Pathogenic Gene Mutation in Human Embryos. *Nature* 548: 413–419.

The National Academies of Sciences, Engineering, and Medicine. 2017. *Human Genome Editing. Science, Ethics and Governance*. Washington, DC: The National Academies Press.

Genome Editing, Non-Identity and the Notion of Harm

Alina Omerbasic

Keywords

Future Generations, Non-Identity Problem, Nonmaleficence, Risk, Suffering

Abstract

New technologies such as CRISPR-Cas9 promise and allow precise and astonishingly simple ways of editing the human genome. They thereby fuel the wish to cure serious genetic diseases by eradicating them where they emerge – in the genome itself. In order to assess this technology from an analytic-philosophical perspective I want to question whether its application violates the principle of nonmaleficence. By pointing out the now famous *Non-Identity Problem* (NIP) Derek Parfit showed that this proves to be very difficult in cases in which our actions affect the *welfare together with the identity* of future living individuals. My interest in this paper is in exploring whether we face the NIP in the case of genome editing, or not. In my view, a two-tier analysis is needed, distinguishing between the question of harm to the "edited-embryo" itself and harm to its descendants in the case of genetic interventions in the germline. While we do not face the NIP in the first generation of genetically edited embryos, and while there is a high risk of harm in these cases, it *does* hold in the next generation, which makes it surprisingly hard to oppose invasions in the human germline with the help of the principle of nonmaleficence. I present and discuss my analysis here, which might help to resolve certain misconceptions of harm and to set boundaries in an apparently harm-free area.

M. Braun et al. (Eds.), *Between Moral Hazard and Legal Uncertainty*, Technikzukünfte, Wissenschaft und Gesellschaft / Futures of Technology, Science and Society, https://doi.org/10.1007/978-3-658-22660-2_5

1 Introduction

The starting point of this paper is the assumption that the *principle of nonmaleficence* is one of our most practical, action-guiding principles. This is the case not only in everyday decision-making, but also in controversial debates such as those that concern the permission of new, and sometimes critically viewed, medical devices or technologies. New technologies, such as CRISPR-Cas9, promise and allow precise and astonishingly "simple" ways of editing the human genome, thereby fueling the desire to cure serious genetic diseases by eradicating them where they emerge – in the genome itself. As promising as this sounds, it has also provoked criticism from the beginning. In both public and academic debate, it is pressed for reasonable risk assessment in the context of genome editing, which comes down to a demand for the consideration of foreseeable harms or losses of affected individuals, such as the edited individuals themselves in the case of genome editing.[1] In order to assess this technology from an analytic-philosophical perspective in the following, I take up this issue and ask whether the use of genome editing technologies like CRISPR on human beings violates the principle of nonmaleficence.

2 The Principle of Nonmaleficence and the Notion of Harm

Now, what does the principle of nonmaleficence concretely amount to? It states that our freedom of action ends where our actions harm or run the risk of harming other people. If we want to apply this principle to the question of the legitimacy of the use of CRISPR-technologies on human beings (whether in somatic cells or the germline), then we have to clarify whether or not its use harms the manipulated individuals.

1 In this article I do not address the question of harm to other persons who might be affected by genome editing of certain individuals (like family members or society in general), for this is another issue. In the end, the question here would be whether or not we would want to bear the consequences of considering harm to third parties in such cases at all, because one might – as some thinkers do – argue for example that children with disabilities should not be born due to the "cost" for others (cf. Rakowski 2002, 1345–1414; this is not exactly what Rakowski argues for, but his conclusions could be intepreted in this way quite easily. However, I find Rakowski's conclusions rather unsatisfactory for reasons that I cannot explain here).

Despite appearances, the notion of harm is anything but simple. Following the legal philosopher Joel Feinberg, we can characterize an act as a "harming" one[2] if it meets the following two requirements: the so-called *worsening condition* and the *counterfactual condition*. Put simply:

- *A*'s act harms *B* if *B*'s interests or welfare are in a worse condition than they were before *A* acted, and in a worse condition than they would be had *A* not acted as she did.

Therefore, a harming act has to, first, make the affected individual *B worse off*. That is to say, the action must violate her interests or rights. And second, *B* has to be worse off than she would have been had *A* not acted as she did. That is the counterfactual element seen in the second clause (cf. Feinberg 1986, 145 ff.).

As plausible as Feinberg's person-affecting notion of harm is, it proves to be very problematic in cases where our actions affect the welfare together with the identity of future living individuals, that is, of people who are *not yet* existent. In such cases, we face the so-called "Non-Identity Problem" (NIP), which became acquainted by the works of Derek Parfit, especially in his book *Reasons and Persons* (Parfit 1984).

In this paper, I want to investigate whether – and if so, in which configurations of cases – we face the NIP in the context of human genome editing. Once this is clear, I want to offer a solution to the NIP by introducing a new action-guiding principle for genuine non-identity cases that can capture their distinctiveness and avoid the errors of previous proposals.

However, before I get to any of that, I will briefly illustrate the NIP based on two examples that, in one way or another, are discussed in the literature. Even though this problem might at first seem to be no more than an exercise in philosophical hair-splitting, it is of great theoretical and practical importance, because its consequences are weighty, not just for everyday decision-making and moral argument, but for legal practice as well.

---

2 As opposed to a *harmful* act or condition for which no one is to blame, like an earthquake, or some other natural disaster.

3 The Non-Identity Problem

Consider the following two examples:

E1: *The impatient woman* (pre-conception harm)
A young woman has decided to have a baby. Her doctor tells her that she has some disease and if she conceives now, her baby will have an irreversible disability – in this case, maybe a mild mental disability. Even if this disability does not render the child's life not worth living, it is going to have a substantially negative impact on her quality of life. The doctor asks the young woman to postpone the pregnancy for another month, because she could thereby prevent her child from having the disability. Informed about the consequences, she nevertheless decides to conceive now, and, as a result, she gives birth to a baby with a mild disability (cf. Parfit 2010; Brock 1995).

E2: *The drinking mother* (post-conception, but prenatal harm)
A woman starts drinking alcohol excessively during her pregnancy, because her husband abandoned her, and she does not know how to handle the prospect of being a single mother in any other way. Although she is fully aware of the bad consequences for her baby, she does not stop drinking, and a couple of months later, due to the alcohol, her baby is born with certain cognitive and behavioral disorders.

If asked to spontaneously assess these cases, most people would presumably condemn the behavior of both mothers as equally wrong, because they consider the resulting children to be harmed. At first appearance, this assessment makes sense. However, there is a certain feature of the mothers' behaviors that makes them *not* equally wrong. This feature – or the problem – consists in the fact that the affected child in E1 could not have been born other than with a mild disability. Had the mother postponed the pregnancy as the doctor had advised her, she would have born a healthy, but *a different* child, since conception would have taken place at a different point in time, namely one month later.

The crucial point here is that we cannot locate harm done to a person in cases where we would intuitively do so. As long as we are asking about the condition *of one and the same individual*, before and after a possibly harming act, we cannot say that it would have been "better for her" had her parents acted differently, or, in cases like these, that the resulting children "have been harmed" by their parents' reproductive decisions. The only possibility to protect *these specific children* from the disabilities consists in not giving birth to them, which seems to be paradoxical.[3]

4 The Pitfalls of Identity

This paradox is based on the fact that, depending on how the future parents actually decide to act, different respectively "non-identical" children will be born. In his book *Reasons and Persons* (Parfit 1984), Parfit summarizes the dependence of the existence and identity of an individual on when, and under which circumstances, it was conceived. He does this in his so-called "Time-Dependence Claim" (TD2), which states: "If any particular person had not been conceived within a month of the time when he was in fact conceived, he would in fact never have existed" (Parfit 1984, 352).

To me, it seems quite generous to link the relevant time frame to the female reproductive cycle, as Parfit apparently does; strictly speaking, it is a matter of milliseconds, because the genetic make-up depends not only on the female egg cell, but also on the sperm that "wins the race". Whatever the case may be, Parfit's claim is, on the basis of biological facts, highly plausible.

It is important to note that what we are focusing on here is not personal identity in a metaphysically charged sense, but mere *numerical identity*. We can consider a born individual to be numerically identical with the embryo it emerged from in the same way as we consider a butterfly to be numerically identical with the caterpillar it emerged from. At this point, nothing has been said about their (obviously not given) *qualitative identity*, or the different states and rights attributed to them at their different levels of existence and development. Likewise, the embryo in E1 is numerically identical with the impaired child to whom the impatient woman gives birth to later; if it weren't, the NIP would, of course, not arise, and we could make a convincing point by telling the impatient woman: "Whether you conceive now or in one month, you will give birth to one and the same individual, and this individual will be better off if she is conceived after you make it through this illness. If you conceive now you are harming her, so wait!" But this – again, in light of the biological facts – is not true. Following the philosopher David Heyd, I will call non-identity cases like E1 "genesis cases". These are cases in which some undesired

3 Gregory Kavka initially called this phenomenon the "future individuals paradox" (Kavka 1982). Later, Derek Parfit called it the "Non-Identity Problem" (Parfit 1984).

conditions or traits are tied to the conception, and therefore, to the existence and identity of a certain individual.[4]

Please note that we encounter the NIP only in such cases. It *does not* arise in cases like E2. These are cases of common post-conception harm, in which the child is made worse off, respectively harmed *after conception* and during pregnancy. Had the pregnant woman not been drinking alcohol in this case, the resulting child would (presumably) not have had to deal with cognitive and behavioral disorders.

What consequences follow from this for the assessment of the mother's behavior in E1? If she did not harm or wrong her child, then – focusing on the principle of nonmaleficence – it is not legitimate to intervene in her reproductive freedom. This is a surprising conclusion, since I, probably along with most people, would condemn her impatience. The same holds, for example, in cases of positive selection for usually rather disadvantageous traits, like deafness or blindness. Let's say a deaf or blind couple wishes to have a deaf or blind child, since they want their child to grow up in a world as they know it. *Preimplantation genetic diagnosis* offers them the possibility to choose and use only "deaf" or "blind zygotes" from a number of fertilized eggs for implantation into the woman's uterus. Now, one might be inclined to say that these parents are harming the resulting child by choosing a sensually impaired zygote, but in fact they are not. If they had chosen a "hearing" or "seeing" one, it would have been *a different* embryo and the procedure would have resulted in *a different* child. In short: Had they acted differently the deaf or blind child concerned would not have been better off, for it would not have been born at all.

Now, does this much debated problem also arise in the context of the use of technologies like CRISPR on human beings? Put another way, can we advance the so-called "Non-Identity Argument" (NIA) in the case of CRISPR-Cas9 (assuming of course that the procedure, as such, is well-elaborated, safe and functions reliably)? If, following the NIA, there is no harm done to the manipulated individual, we cannot say that she was harmed by the application of the procedure, nor that, in consequence, it has to be restricted or condemned.

5 Do We Face the Non-Identity Problem in the Case of CRISPR?

Bearing in mind both the banal combination of circumstances that led to *our* existence, and Parfit's TD2, we should recognize the need for a careful consideration

4 David Heyd calls them "genesis problems", cf. Heyd (1992, xii); Brock (1995, 272).

of the actual cases we are dealing with in the context of genome editing on human beings. In my view, what is needed is a *two-tier analysis* that distinguishes between the question of harm to the very first edited embryo itself, let's say "embryo zero" (whether it involves somatic or germline cells), and harm to its descendants in the case of germline intervention.

5.1 Considering the Case of Embryo Zero

Let's have a look at embryo zero and genetic intervention via CRISPR first. Is embryo zero harmed by the manipulation, or not? To begin with, in my view, we are *not* dealing with a genuine non-identity case here, since the embryo that is about to be manipulated is already existent, and its identity is already determined in the sense of Parfit's TD2 (putting aside twinning[5]). In short, *should* something go wrong during or because of the gene manipulation – whether it is done for the purpose of treatment or enhancement –, it is safe to say that there is harm done to the individual concerned, because both the worsening condition and the counterfactual condition of our common person-affecting principle of harm are met. Moreover, if there are certain undesirable off-target effects to be expected, the procedure could or should be restricted if not prohibited. If all of this is not the case, gene manipulation can be legitimate in the case of embryo zero.

5.1.1 A Needed Distinction of Cases

Even though this case could be considered plain sailing, a clever critic might note: "Well, even if something goes wrong, the individual concerned has not been harmed in terms of being made worse off by her parents' decision to undertake the manipulation. Had her parents acted differently, she wouldn't be better off, she wouldn't exist at all. Remember that they didn't want to have a child in the first place; they only wanted the child if they were allowed to perform the intervention. Now, who would say that the affected individual would be better off nonexistent, just because the procedure had some off-target effects?" Here, the genetic intervention is taken to be a "prerequisite" for the embryo's existence, respectively the parents' decision to have a child in the first place. A look at the vast literature dealing with the NIP shows that this line of thinking is not uncommon. However that may be, I am of the opinion that a more fine-grained analysis is needed as well as a clear distinction

5 The only oddity twinning involves is that we would probably, as a result, face the NIP "two times".

between the cases discussed in the literature. In fact, there are many cases discussed that are not non-identity cases at all.[6]

In my view, considering the manipulation as a prerequisite for having a child in the first place does *not* make it a genuine non-identity case, in which we are forced to accept the no-harm-done result. We are dealing with a simple case of "post-conception manipulation" here. Put differently, once this specific individual is conceived, the parents can, despite their prior plans, simply refrain from manipulating her. We would still be dealing with the *very same individual* in the sense relevant to Parfit's TD2 (again, putting aside twinning). All in all, our common notion of harm applies to this case, and there is no need to accept the alleged no-harm-done result if anything goes wrong.[7] The parents' decision to manipulate the embryo constitutes harm in case something goes wrong, because this specific individual would have been better off not being badly manipulated.

5.2 Considering the Descendants of Embryo Zero

Let us consider the question of harm to the descendants of embryo zero in the case of genetic intervention in the germline. We are confronted with a different situation here. Consider that the individuals affected were not conceived at the moment the decision to undertake the germline invasion in embryo zero was made. Consequentially, that decision had "identity-determining effects" in what I call the "genesis sense", meaning that the descendants of the edited embryo zero literally owe their existence to the decision that led to their existence – including the decision to perform the genetic manipulation of embryo zero in the first place.[8] If

6 For example Gregory Kavka's infamous "slave child example" (cf. Kavka 1982, 100). For more details, see my forthcoming dissertation.

7 It is another question, of course, whether there is any ethical problem if we can predict or assume the affected individual's consent to the procedure, and everything goes well, as it might be the case in the therapeutic use of gene-editing technologies. To me, it seems difficult to say what speaks against their application in this case. A prohibition could at least not be justified by harm, for who regrets that he was not born with muscular dystrophy, or even with enhanced imagination or health, for example?

8 I intend to further differentiate between "identity-determining effects in the genesis sense" and "identity-determining effects in the accidental sense". This differentiation is based on another differentiation, namely the one between essential and accidental features of an individual; while essential features are "to the people", accidental features seem "added", just secondary. Consider for example the decision to sell a baby as a slave (see Gregory Kavka's slave child example [cf. Kavka 1982, 100]). This undoubtedly questionable decision has identity-determining effects in the mere accidental sense, for

her germline had not been edited, her life would have taken a different course, and consequentially, different descendants would have been born (she would presumably have met a different partner or at least they would have procreated at another point in time, and consequentially would have conceived different children etc.).

To sum up, we face the NIP with regard to the welfare of the descendants of embryo zero, which makes it very hard to argue against genetic interventions in the germline with the help of the principle of nonmaleficence. No matter what state the descendants of embryo zero are born into, as long as their existence is not *worse than nonexistence* there is – following the NIA – no harm done to them by the fact that they are descendants of a manipulated individual. The only "risk" the descendants face is the risk of their existence, while the alternative for them is their non-existence, which logically cannot be considered as harmful to them. In my opinion, this is a surprising result, insofar as genetic interventions in the germline are usually (and, intuitively, quite rightly) regarded as especially problematic – so problematic that even scientists have, for the time being, agreed on a moratorium, and the German Ethics Council is calling for a global-political discourse and international regulations in its latest ad hoc recommendation (cf. German Ethics Council 2017).

6 How to Deal with the Non-Identity Problem?

Parfit's publication on the NIP triggered an extensive debate on the question of how to deal with this problem, but finding a solution is far more difficult than one might think. I will not review this discussion here, but will instead briefly introduce my own proposal for a solution to genuine non-identity cases, with reference to Parfit himself. In my view, it makes no sense to ask about the worsening of welfare of a certain individual by an action, if the individual, whose welfare we are trying to protect, is, as a result of the alternative course of action, non-existent. That is to say, it does not make sense to apply our common person-affecting notion of harm in genuine non-identity cases. What basically happens when we apply our common person-affecting notion of harm in these cases is that we thereby eliminate the

selling the child constitutes merely its "identity as a slave"; "merely", because being born a slave, in my opinion, is something "supervene", or at most an "accidental" feature, and different from having a certain genetic make-up. In my view, only actions with identity determining effects in the genesis sense lead to the Non-Identity Problem. For more details concerning this distinction see my forthcoming dissertation and a forthcoming article concerning the question of harm in the case of embryo donation.

basis for every kind of "outcome-oriented" comparison or assessment of an action, which proves that our common person-affecting harm principle, like Feinberg's, is simply the wrong "tool" in this context.

However, this situation is comparable to the attempt to open a lock with a spoon. Just because a spoon does not help to open a lock, it does not mean that we therefore have to discard spoons altogether. What it does show, though, is that spoons are the wrong tools to use for this job. What is needed, therefore, is an action-guiding principle for non-identity cases that can capture their distinctiveness. Following Derek Parfit and Dan Brock (Buchanan et al. 2000), in what follows I argue that a promising, but also controversial solution consists in replacing, or rather *supplementing*, our common person-affecting harm principle with a so-called "non-person-affecting harm principle" (NPAHP) in non-identity cases. Such a principle seems better suited to capture the moral wrong done in these cases, for the moral wrong does not consist in the reduction of a particular individual's well-being. Instead, we are dealing with another *category of harm* in non-identity cases, which is best captured by a consequentialist NPAHP.

Unlike our common person-affecting harm principle, Parfit's and Brock's proposed principles do not require that the *very same* individual who experiences serious suffering and limited opportunity in one course of action needs to live and be free from these burdens in the alternative course of action. Therefore, the proposed principles cope with the nature of "genesis", or non-identity cases, insofar as they assess two options of action independently of their impact on one specific affected individual.

Since the general idea of NPAHPs is fairly controversial, and the proposed principles of Brock *(N)*[9] and Parfit *(Q)*[10] are not elaborated in detail, their proposals are still somewhat underestimated, if not misunderstood, in the literature. Instead of clarifying this subject at this point, I want to introduce my own proposal, namely

9 Brock's *Non-person affecting harm principle (N)*: "Individuals are morally required not to let any child or other dependent person for whose welfare they are responsible experience serious suffering or limited opportunity or serious loss of happiness or good, if they can act so that, without affecting the number of persons who will exist and without imposing substantial burdens or costs or loss of benefits on themselves or others, no child or other dependent person for whose welfare they are responsible will experience serious suffering or limited opportunity or serious loss of happiness or good." (Buchanan et al. 2000, 249)

10 Parfit's *Same Number Quality Claim (Q)*: "If in either of two possible outcomes the same number of people would ever live, it would be worse if those who live are worse off, or have a lower quality of life, than those who would have lived." (Parfit 1984, 360)

a refined version of a NPAHP. This principle has some advantages over Brock's and Parfit's proposals, which I shall explain elsewhere.[11] I call this principle, *N**:

- If we are to choose between two actions and our choice has identity-determining effects in the genesis (or accidental) sense, it is wrong to choose the action by which, consequentially, any future living individual, for whose existence and welfare we are responsible, experiences serious suffering and limited opportunity, if we can prevent that without imposing substantial burdens or cost on ourselves, themselves or others.

6.1 *N** and the Question of a Justifiable Use of Technologies like CRISPR-Cas9

Confronted with the question of whether or not to use CRISPR-Cas9 for a germline intervention, we do, in fact, face a choice that has identity-determining effects in the genesis sense and that results in the existence of individuals for whose existence and welfare we are responsible. However, the choice to actually undertake such an intervention with the help of CRISPR does not necessarily lead to suffering and limited opportunity. Consequently, *N** does not make the use of CRISPR or other artificial reproductive technologies, like selection in the context of preimplantation genetic diagnosis, morally unacceptable. To the contrary, following *N**, the application of such technologies might even be considered as an option to prevent suffering and limited opportunity.

Of course, this might lead critics to draw a further conclusion and assume that – as a result of the application of principles like *N** – there is an *obligation* to use technologies like CRISPR. Proponents of this objection might suggest that it would, in the light of our obligation to prevent suffering and limited opportunity, be almost "too risky" to conceive naturally, and that we should therefore use these technologies to ensure at least a favorable starting point for the resulting children. Now, should we, following *N**, prevent suffering and limited opportunity by increasing or enhancing physical or cognitive capacities of our descendants? Do we need to "breed optimally", and produce "the best children possible"?

I don't think so, and the reason is quite simple: Being "the best" is not just relative to other individuals, but is also relative to a certain environment; put differently, "the

11 I embrace the idea that a clearly formulated non-person-affecting harm principle can be defended against a number of objections raised in the literature, and can solve many of the commonly discussed test cases. For more details see my forthcoming dissertation.

best" is always "the best for". Now, if parents choose or create "the best child", or "the perfect child", they are *not* ensuring that it will not undergo suffering and limited opportunity. For they can obviously quite seriously restrict their children's scope for actions, or creative leeway, through specific preferences regarding the capacities of their children. This leads to the conclusion that parents should not have a maximal freedom of choice in this area. The basic problem is that the resulting individuals might turn out to be very inflexible with respect to their own life plans. Having the best traits in one context might be disadvantageous in another, which speaks against an obligation to enhance in the context of a consequentialist principle like *N**.

There is also a reason why we should not be guided by society's convictions or social prejudices in the context of selection or genetic manipulation in general, which is captured by *N**. Following *N**, we should not be guided by criteria or traits that are desirable in a certain society (following clichés, à la "intelligence is good", "beauty is good", or "homosexuality is disadvantageous"). The additional problem here is that – apart from making the resulting individuals inflexible for their own life plans – we do not know how any given society will develop. Consequently, it would, in terms of *N**, not be very "clever" to exclusively select for certain traits that are desirable in our current society; these convictions and preferences simply won't stand the test of time which should be considered, especially in the context of germline intervention.

In sum, *N** does not necessary exclude the use of technologies like CRISPR in the case of interventions in the germline in order to prevent suffering and limited opportunity. However, it allows them only *within limits,* for otherwise, their use would contradict the principle. The resulting individuals should be flexible and diverse enough to choose the life they want to pursue. In practice, this means that people should, for example, not *manipulate towards,* or *select for* some ideal of beauty or specialized skills or traits. While increased imagination or intelligence (only up to a certain extent) might be desirable in any conceivable environment, impaired senses like deafness or blindness are not. Following principles like *N** selecting or manipulating technologies like CRISPR-Cas9 are to be understood as "options", not as "ends in themselves". NPAHPs do *not prohibit*, but they do *constrain* the application of such technologies, by their focus on the criteria of flexibility and diversity.

7 Concluding Remarks

I started by questioning whether the application of technologies like CRISPR on human beings violates the principle of nonmaleficence, because, if it does, we would have a perfect reason to restrict or even prohibit their use. We saw that the wide-spread person-affecting notion of harm that underlies the principle of non-maleficence is not capable of dealing with so-called genesis cases, i. e., actions that affect the welfare together with the identity of future living individuals. In such cases, we face the so-called Non-Identity Problem, respectively the Non-Identity Argument can be applied to show that there is no harm done, no matter what state the individuals concerned are born into.

What I demonstrated is that, *first*, the Non-Identity Argument does not hold in the first generation of genetically edited embryos, since we are dealing with simple cases of post-conception manipulation here, where there is a high risk of harm; and *second*, that it does hold for the next generations, which makes it surprisingly hard to argue against the use of genome editing technologies like CRISPR for intervention in the germline. I then presented my solution to the Non-Identity Problem by introducing a refined action-guiding principle for non-identity cases that can capture their distinctiveness and avoid the errors of previous proposals.

All in all, technologies like CRISPR-Cas9 are doubtlessly promising, but they should be treated with caution. However, my analysis showed that we can hardly speak of "impending risks", or "harms", in the conventional sense of our common notion of harm, especially in the case of the usually mostly feared germline intervention. I hope to have shown that we need different patterns of argumentation in genuine non-identity cases, and that non-person-affecting harm principles like N^* offer a reasonable way to deal with such non-identity cases. We should keep in mind that the aim of such principles is to make sure that future living individuals do not fall outside of the scope of ethical consideration and to provide them with as much room as possible with regard to how they want to live their lives. Even though such principles cannot prohibit the use of technologies like CRISPR-Cas9, they can help us to set boundaries in an allegedly harm-free area.

References

Boonin, David. 2008. How to Solve the Non-Identity Problem. *Public Affairs Quarterly* 22: 127–157.
Boonin, David. 2014. *The Non-Identity Problem and the Ethics of Future People*. New York: Oxford University Press.
Brock, Dan W. 1995. The Non-Identity Problem and Genetic Harms. The Case of Wrongful Handicaps. *Bioethics* 9: 269–275.
Brock, Dan W. 2005. Shaping Future Children. Parental Rights and Societal Interests. *Journal of Political Philosophy* 13: 377–398.
Buchanan, Allen, et al. (eds.). 2000. *From Chance to Choice. Genetics and Justice*. Cambridge: Cambridge University Press.
Feinberg, Joel. 1984. *Harm to Others. The Moral Limits of the Criminal Law*. New York: Oxford University Press.
Feinberg, Joel. 1985. Wrongful Conception and the Right not to be Harmed. *Harvard Journal of Law and Public Policy* 8: 57–77.
Feinberg, Joel. 1986. Wrongful Life and the Counterfactual Element in Harming. *Social Philosophy and Policy* 4: 145–178.
Feinberg, Joel. 1988. *Harmless Wrongdoing. The Moral Limits of the Criminal Law*. New York: Oxford University Press.
Feinberg, Joel. 1992. The Child's Right to an Open Future. In *Freedom and Fulfillment. Philosophical Essays*, Feinberg, Joel (ed.), 76–97. West Sussex: Princeton University Press.
German Ethics Council. 2017. *Germline Intervention in the Human Embryo: German Ethics Council Calls for Global Political Debate and International Regulation. Ad Hoc Recommendation*. http://www.ethikrat.org/publications/ad-hoc-recommendations/files/recommendation-germline-intervention-in-the-human-embryo.pdf. Accessed 15 February 2018.
Glover, Jonathan. 2006. *Choosing Children. Genes, Disability, and Design*. Oxford: Clarendon Press.
Hanser, Matthew. 1990. Harming Future People. *Philosophy and Public Affairs* 19: 47–70.
Harman, Elizabeth. 2004. Can We Harm and Benefit in Creating? *Philosophical Perspectives* 18: 89–113.
Heyd, David. 1992. *Genethics. Moral Issues in the Creation of People*. Berkeley: University of California Press.
Heyd, David. 2014. Parfit on the Non-identity Problem, Again. *Law & Ethics of Human Rights* 8: 1–20.
Holtug, Nils. 1999a. Does Justice Require Medical Enhancements. *Journal of Medical Ethics* 25: 37–143.
Holtug, Nils. 1999b. Utility, Priority and Possible People. *Utilitas* 11: 16–36.
Holtug, Nils. 2002. The Harm Principle. *Ethical Theory and Moral Practice* 5: 357–389.
Kavka, Gregory S. 1982. The Paradox of Future Individuals. *Philosophy and Public Affairs* 11: 93–112.
Kitcher, Philip. 1997. *The Lives to Come. The Genetic Revolution and Human Possibilities*. New York: Touchstone.
Parfit, Derek. 1982. Future Generations: Further Problems. *Philosophy and Public Affairs* 11: 113–172.

Parfit, Derek. 1984. *Reasons and Persons*. Oxford: Clarendon Press.

Parfit, Derek. 2004. Overpopulation and the Quality of Life. In *The Repugnant Conclusion. Essays on Population Ethics*, Ryberg, Jesper, and Torbjörn Tännsjö (eds.), 7–22. Dordrecht: Springer.

Parfit, Derek. 2010. On Doing the Best for Our Children. In *Population and Political Theory*, Fishkin, James S., and Robert E. Goodin (eds.), 68–80. West Sussex: John Wiley & Sons.

Parfit, Derek. 2011. *On What Matters*. Oxford: Oxford University Press.

Rakowski, Eric. 2002. Who Should Pay for Bad Genes? *California Law Review* 90: 1345–1414.

Roberts, Mary A. 2004. Person-Based Consequentialism and the Procreation Obligation. In *The Repugnant Conclusion. Essays on Population Ethics*, Ryberg, Jesper, and Torbjörn Tännsjö (eds.), 99–127. Dordrecht: Springer.

Roberts, Mary A. 2007. The Non-Identity Fallacy: Harm, Probability and Another Look at Parfit's Depletion Example. *Utilitas* 19: 267–311.

Roberts, Mary A. 2009. What is the Wrong of Wrongful Disability. From Chance to Choice to Harms to Persons. *Law and Philosophy* 28: 1–57.

Roberts, Mary A. 2009. The Nonidentity Problem and the Two Envelope Problem: When is One Act Better for a Person than Another? In *Harming Future Persons. Ethics, Genetics and the Nonidentity Problem*, Roberts, Melinda A., and David T. Wassermann (eds.), 201–228. New York: Springer.

Roberts, Mary A., and David T. Wasserman (eds.). 2009. *Harming Future Persons. Ethics, Genetics and the Nonidentity Problem*. New York: Springer.

Robertson, John A. 1994. *Children of Choice. Freedom and the New Reproductive Technologies*. Princeton/New Jersey: Princeton University Press.

Robertson, John A. 1996. Genetic Selection of Offspring Characteristics. *Boston University Law Review* 76: 421–482.

Robertson, John A. 2003a. Procreative Liberty in the Era of Genomics. *American Journal of Law & Medicine* 29: 439–487.

Robertson, John A. 2003b. Extending Preimplantation Genetic Diagnosis: The Ethical Debate. *Human Reproduction* 18: 465–471.

Robertson, John A. 2004. Procreative Liberty and Harm to Offspring in Assisted Reproduction. *American Journal of Law & Medicine* 30: 7–40.

Savulescu, Julian. 2001. Procreative Beneficence. Why We Should Select the Best Children. *Bioethics* 15: 413–426.

Savulescu, Julian, and Guy Kahane. 2009. The Moral Obligation to Create Children with the Best Chance of the Best Life. *Bioethics* 23: 274–290.

Genome Editing and the Transgression of the Species Boundary

Does Species Affiliation Have an Ethical Relevance?

Markus Rothhaar

Keywords

Gene Editing, Hybrids, Interspecies Beings, Life Forms, Naturalness, Neo-Aristotelianism, Personhood, Species Argument, Transgenic Organisms.

Abstract

One of the main fields of research in genome editing is the manufacture of transgenic organisms. If this includes genetically human components, then we must not only ask whether such techniques should be allowed, but also what the moral status of "cross species beings" with a genetically "human" component would be and whether the possibility to manufacture them affects the validity of bioethical arguments from species affiliation. In my chapter I want to show that the two latter questions are less dramatic than it might seem, because most, if not all transgenic organisms are not real "interspecies beings", but rather belong to one of the original species, or constitute a new species. As for the first question, possible epistemic uncertainties about the status of transgenic organisms call for a precautionary approach.

M. Braun et al. (Eds.), *Between Moral Hazard and Legal Uncertainty*, Technikzukünfte, Wissenschaft und Gesellschaft / Futures of Technology, Science and Society, https://doi.org/10.1007/978-3-658-22660-2_6

1 Introduction

In January 2017, the German press reported (Müller-Jung 2017) that it is now purportedly possible to manufacture "cross-species beings from pigs and humans". In the background, behind these reports is one of the most important current research fields in biomedicine. It is here that stem-cell research and genome editing collaborate in the attempt to grow genetically "human" organs in animal organisms (in this particular case, in domesticated pigs). While the reports – with the exception of their pithy headlines – were quite neutral by German standards, emotions ran high in the commentary sections of the newspapers and magazines. This manifested itself in an almost unanimous and, in part, extremely emotional opprobrium, the tenor of which was something like: "Now you (that is to say, you scientists), have finally gone too far. This is something that can never be accepted". These vehement, and remarkably unanimous, reactions show that transgressing the species boundary is clearly perceived by many people, regardless of their other views, their religious and cultural backgrounds etc., as breaking a massive taboo. Hence, the species boundary obviously plays a crucial role in *human self-understanding*, to which an enormous ethical and – where human self-understanding is concerned – hermeneutical relevance is imputed. At the same time, there is a wide-spread opinion among professional philosophers that it is not possible to rationally substantiate the ethical relevance of the boundary of species. The most prominent objection to the assumption of such relevance is that of an unjustified "speciesism". Yet, one need not be a proponent of this pointed "speciesism-critique" to have a problem with the assumed ethical relevance of species boundaries. There seems to be a considerable gulf between the professional philosophical debate and the widespread perceptions of the subject in society.

Yet, why is it now, in the context of genome editing, that the problem of species boundaries comes into focus? Is it merely accidental that the boundaries of species can be transgressed by means of genome editing, a transgression which, in itself, only constitutes a small part of genome editing? To me, this seems not at all to be the case. Rather, the transgression, on the genetic level, of the boundaries between species is one of the important endeavors in genome editing. Apart from the intentional deactivation (or "silencing") and activation of genes, a considerable part of genome editing consists in the creation of transgenic organisms. Surely, this might eventually change, namely when synthetic genes can be manufactured – genes that do not yet exist in nature and are not merely synthetic copies of their already existing counterparts. At present, however, there exists a conceivable and in fact intimate connection between genome editing and the problematic of species.

Therefore, the expression, "the transgression of species boundaries", might be said to gain its substantial meaning in the creation of transgenic organisms.

The expression itself, however, is in fact ambiguous, since the "transgression of species boundaries" might also refer to the creation of those beings envisioned by the so-called "transhumanists". One might understand this program "merely" in terms of a recombination of human DNA with the DNA of animals or plants: that is, in terms of the aforementioned manufacture of transgenic organisms. "Transgression", however, can also connote something else here, namely an overcoming of the (purported) constraints that nature has imposed on us: the surmounting of the natural pre-givenness of our bodies and their properties in favor of a free self-formation of the same. This special characteristic of transhumanism simultaneously raises political and legal questions that are not implicated in the mere notion of the "creation of transgenic beings". This, however, is not the theme of this chapter. Nevertheless, it ought to be mentioned for the sake of completeness.

2 Ethical Questions Regarding the Manufacture of Transgenic Organisms

In the main part of this chapter, I bring into view the respect in which the "transgression of the species boundary" might raise ethical questions where the manufacture of transgenic beings is concerned. As I see it, two different dimensions of the question can be distinguished here: firstly, that of "naturalness" as such, and secondly, that of affiliation to the human species.

Those who consider "naturalness" to be a value in itself, or a normatively relevant category as such, might come to the conclusion, that in a general sense, the creation of transgenic organisms contravenes an ethical norm. In the philosophical discourse, however, such considerations are largely rejected as, at best, relics of religious beliefs that cannot serve as the basis of legislation in a liberal, pluralistic and constitutional state. Yet, even if "naturalness" is *not* generally considered to be a normatively relevant category, it might nevertheless be the case that affiliation to the natural species "human" possesses a normative relevance. As the aforementioned reactions to the manufacture of human-pig hybrids have shown, the transgression of the boundaries between human and animal beings manifestly touches on a deeply rooted taboo, displaying serious implications for human self-understanding. However, on closer inspection, a variety of different dimensions of interrelating ethical issues can be identified:

a. the question of the *permissibility* of the manufacture of human-animal composite beings in general;
b. the question of the *moral status* of such beings, once manufactured;
c. the question of the *validity* of the so-called "argument from species" in bioethics, i.e., the argument that affiliation to the species "human" is either a basis for the granting of human dignity and human rights or, and at the very least, the criterion for granting human dignity and human rights. It is thus a question of whether or not the arguments from species can be upheld at all if the affiliation to species is no longer something naturally given, but becomes itself an artefact.

Points c) and b) in particular are intimately interrelated. For, if species affiliation is no longer something "natural", then it is of crucial importance for the moral status of such beings whether or not affiliation to the human species can still serve as a basis or criterion for the granting of human dignity and human rights. Thus, the last question refers fundamentally back to the normative significance of affiliation to the human species, and in this respect, provides a good starting point for approaching the two ancillary questions. Therefore, I begin with the latter point. For by means of it, we can come to understand both of the other questions, and in particular, gain important insights regarding the first question. The problem that is concealed here ultimately points to the fundamental relationship between human nature and normativity.

3 The Bioethical Argument from Species

To my mind, the argument from species raises *two* variations that are philosophically feasible (I omit the infeasible ones here):

1. that to which Robert Spaemann refers as the "transcendental-pragmatic" variant;
2. the (in the broadest sense of the word) (Neo-)Aristotelian variant, that refers to species-typical forms of life.

3.1 The "Transcendental-Pragmatic" Argument According to Robert Spaemann

Spaemann begins by arguing that the idea of human dignity and human rights implies a form of unconditionality that makes placing them under the condition of the existence of particular properties appear inappropriate. Such a "conditional granting" therefore misunderstands the substance of human rights and human dignity, in that their possession does not rest upon a conditioned claim. Spaemann's trenchant first formulation of this discrepancy appears in his book, "Happiness and Benevolence", in which he argues:

> "If there is to be anything like human rights, then they are only possible given the presupposition that no one is entitled to render judgment whether someone is a subject of such rights. For the logic of human rights entails that the human is not a member of the human society who is co-opted on the basis of certain qualifications; rather that each enters into the society on the strength of one's own claim. But on the strength of one's own claim can only mean: on the basis of belonging biologically to species homo sapiens. Any other criterion would make the one a judge over the others. Human society would become a closed shop, and the thought of human rights would be undone at its very basis. Only when the human is recognized as a person on the basis of that which he is or she is by nature does the recognition apply to the self and not to him or her as someone who fulfils a criterion, which others have established for their recognition." (Spaemann 2000, 170–171)

These reflections are often understood as a kind of elaborated variation of tutiorism, that is to say, of all of the criteria which could be applied to determine the boundaries of the community of rights, that one should be chosen which is the most general and the least arbitrary: the affiliation to species. Walter Schweidler, for example, reads Spaemann in such a way when he interprets these reflections in the sense of a normative claim not to deny the staus of a bearer of subjective rights:

> "If we understand 'human dignity' as the placing of a legal concept at the peak of the legitimatory pyramid of the order of our community, then we are not committing ourselves to a material definition from which it can be deduced which beings do or do not befit this dignity, rather we are committing ourselves to one of our acts of delimitation which abandons the legal order: namely, to the basic proscription for all ethically responsible action, the judging of other members of the association of right as to whether or not they are befitting of dignity." (Schweidler 2008, 177)

However, if this were the point, then the argument would be circular. This becomes clear if in the last sentence we replace the concept "dignity" (in accordance with Schweidler's own definition of the concept) with the "status of a bearer of subjective rights". The claim in this case would merely be that members of the community of rights may not exclude other members of the community of rights from it. Yet, this only makes sense if the latter already *have* a claim to being treated as members. However, it is exactly this assumption that is controversial where unborn human beings or cross-species entities are concerned.

Yet, perhaps Spaemann can be read differently. The most crucial aspect of the argument may be the reference to human nature, that is to say, the reference to the notion that, in order to be valid at all as an act of unconditioned recognition, entry into the community of rights must result "on the strength of one's own claim", and thus, by virtue of one's own nature. Spaemann obviously invokes the concept of species affiliation in order to operationalize a certain aspect of recognition as a bearer of human dignity: The act of recognition is only such when it does not owe itself to particular, subjective positing, that is to say, when it has not been generated by us as finite subjects in relation to our particular purposes. For that which has not been "made" by us as single, finite subjects, for that which has not been posited by us, but which produces itself, we have always had the concept of nature, or rather, of life. The recourse to nature, or to life, results from the insight that the intersubjective characteristics of the process of recognition, which exceed every individual subject, necessitate thinking of it as independent of all subjective and particular positing.

At this point, it becomes clear that the recourse to the naturalness of species affiliation in Spaemann's argument has precisely the function of guaranteeing the indisposability of human dignity. This itself obviously becomes questionable if species affiliation is no longer something given in a "natural way", but becomes an artefact to be manipulated and manufactured.

3.2 The Neo-Aristotelian Argument from Species

The basic thought of such theories is that every being belongs to a particular species which exhibits its own particular form of life typical to the species in question. This form of life is teleological in character: A being is oriented toward its form of life as towards a *telos* to which (if we are talking about living beings) two characteristics apply:

a. the individual can realize the *telos* more or less successfully;
b. a being realizes its *telos* by itself and brings it forth; it thus exhibits an entelechy.

From this immanent entelechy it can be discerned what this being is according to its essence, according to its *ousia*. The entelechy of human beings is, in turn, oriented toward theoretical and practical rationality, i.e., personality. Accordingly, personhood and "non-personhood" – and this is the crucial point of this position – are not themselves sortal concepts according to which beings can be categorized. "Personhood", as Spaemann says, is not a "natural kind" (Spaemann

2006, 31). Rather, natural kinds (for example, human beings) are persons, while others (like lions or bees) are not. "Personhood" then consists *in a certain way* of being what one naturally is: namely, standing at a reflexive distance, indicated by self-consciousness, to one's own nature. It is this reflexive distance which grounds theoretical and practical rationality.

According to this concept, every affiliate of the species "human being" belongs to a species, the form of life of which, typical to the species, is characterized by theoretical and practical rationality. Furthermore, it belongs to this form of life that those affiliated to the species typically manifest these characteristics at some time in the course of their lives: That is to say, that they go through a process of development. For this reason, they are not excluded from their specific form of life in those phases of life in which they do not exhibit self-consciousness. Rather, it belongs to their form of life that they go through different phases of life, some of which are not characterized by self-consciousness.

In addition, this element of rationality and personhood necessarily "colors" the other moments and aspects of this form of life (for example, sociality, human diet, human sexuality and reproduction, etc.) with the specifically human form of life. For a social being, reason means something other than what it means for a non-social being (if there can even be non-social, rational beings at all). In a rational being, sociality (and sexuality/reproduction, diet, etc.) takes another form, with another significance, than in a non-rational being, as can be observed when the specific sociality of social animals like ants or wolves is compared with the specific sociality of human beings.

Accordingly, the development of a human being from a zygote to the embryonic and fetal stages, to childhood and adolescence is, from the very beginning, something other than among other species. It possesses another significance and another framework. Above all, from the very beginning, it is integrated into a space of reciprocal reflexive recognition, without which we would not have our self-consciousness and therefore no rationality (as, to my mind, Fichte [2000], and in particular, Hegel[1] have trenchantly demonstrated, and which has also been suggested in present developmental psychology – see, for example, the work of Wolfgang Prinz [2012]). In comparison, the natural development of, for instance, a mouse, a cow or even a tree is not integrated into such a space of recognition. At this point, the Neo-Aristotelian and the transcendental-pragmatic argument manifestly interlace.

In this respect, practical rationality is not absent in the phases of life of unborn or new-born, nor is it absent in sufferers of dementia. Rather than being absent,

1 Cf. the famous Master-Slave-Chapter in the *Phenomenology of Spirit* (Hegel 1977).

it "merely" exhibits a different character, in which human sociality (which, as such, is molded by practical rationality, which it in turn conditions) stands in the foreground. This also means that, in the course of the natural development of human life, realized personhood reveals that every living being of the natural kind "human being" is always, in every phase of life, to be addressed as a rational being. For the rationality of such a being is what makes its form of life typical to its species. "Species-specific forms of life" are always only diachronically conceivable.

What are the problems such an approach might lead to, if the limits of species could be genetically transgressed? Firstly, according to my reconstructed argument, personhood is understood as the *telos* and way of being (the *ousia*) of living beings of the natural kind "human being". In this variation of the species argument, the concepts of *ousia* and *telos* thus appear manifestly to require the existence of natural kinds. Yet, is talk of "natural kinds" (apart from its philosophical problematic) at all still plausible, if the boundaries of species can be transgressed by biotechnology? Certain problems result here regarding the tentative normative relevance of human species affiliation. *Other* kinds, whose way of being is also personhood, would be no problem for the theory and would be easily integrated into it (for instance, extra-terrestrial intelligences, possibly also higher primates, etc.). But what about living beings that cannot be considered affiliates of a given species, but which nevertheless exhibit the entelechy of rationality? Would they have to be excluded from the community of rights, since they do not belong to a certain natural kind?

4 The Normative Status of Transgenic Organisms with a "Human Component"

Against this background, it becomes clear why we must ask what kind of normative status is to be attributed to transgenic life forms with a human "component". If my reconstruction of the bioethical argument from species is correct, then the status of human rights and human dignity answers an unconditioned claim proceeding from a living being whose *form of life* is denoted by its personality. Furthermore, it is only persons who are capable of perceiving and answering this claim – as many different authors (such as Kant, Fichte, Spaemann and Lévinas) would well agree, that it is precisely this ability that first constitutes and realizes personality or subjectivity. Correspondingly, the relationship to other persons is always also a relationship to one's own form of life. The answer to the unconditioned claim links itself, it would seem, to affiliation to the species "human" with its specific entelechy: the bringing forth, in the course of its life, of personality. Therefore, affiliates of the species "human", who do not yet or no longer realize their personality, are not excluded from recognition. They remain present in the genealogical and social context of a species whose form of life is essentially characterized by personality. It is at least by virtue of their affiliation to the species "human", and hence, the accompanying genealogical and social context with other affiliates of the species, that they remain within the framework of the relations of recognition.

This becomes problematic, however, if both problems are given at once: If, by reason of genetic manipulation, a living being is denied the possibility to realize itself as a person in the course of its development, and at the same time, by reason of the (same) genetic manipulation, its affiliation to the "human" species (or indeed any other that is characterized by personality) comes into question. The point is that this affiliation must in fact be questionable if this problem is to emerge. If it is *unquestionably* not (or "no longer") a living being of a species characterized by personality, and if this living being *indubitably* could never realize a personality, then that to which we are referring is surely not the bearer of human rights and human dignity. But if it remains unclear, despite genetically induced "impairment", whether we are talking about a (consequently impaired) specimen of the species "human", or a specimen of another species, or no "species" at all, then the question of its status remains open. It seems evident to me that the questions hanging in the air can only be answered if we first take a step back and put the concept of species itself philosophically, scientifically, and ontologically to the test.

5 The Concept of Species Takes the Stand

As is well known, biology admits of a number of differing, sometimes partially contradictory[2], conceptions of species[3]. Among these variants, however, one can discern three different, basic ways of understanding "species". First, there is the morphological understanding of the concept of species: According to this concept, species are groups of living beings which exhibit common morphological traits. In this way, of course, no single morphological trait is adduced that is determinate of a species. Rather, an entire cluster of traits is ordered into a classificatory hierarchy. The classic example of such a system of classification is the Linnaean taxonomy.

2 This shows that the concept of species is not a biological, but a hermeneutical, if not an ontological, one. In any case, it is certainly not, *sensu stricto*, a scientific concept (as little as is the concept of "life").

3 For an overview of the current philosophical debate cf. Richards 2010.

The second basic option for understanding and utilizing the concept of species, Ernst Mayr's population-genetic concept of species (Mayr 1963), is rather more recent. According to this concept, species are defined as groups of living beings that are actually or potentially capable of propagating themselves, that are, *as* groups, "reproductively isolated" from each other. The concept of "reproductive isolation" understands only those creatures as belonging to *one* species, that are capable of reproduction, and the offspring of which are themselves capable of further reproduction (excluding thereby the infamous example of the mule).

In addition to these two definitions of species (which may lead to partly different conclusions), there may be a third definition, which relies on certain ideas from contemporary Neo-Aristotelian approaches, particularly that of Michael Thompson. To my mind, these Neo-Aristotelian approaches exhibit not only a great measure of philosophical and bioscientific plausibility, but also seem capable of integrating the morphological and population-genetic concepts of species into an overarching perspective. The core of this Neo-Aristotelian conception of species is the aforementioned concept of the "form of life". Every living being exhibits a particular *form of life* that permits it, as Thompson says, to be described in terms of "natural-historical judgments" or "Aristotelian categoricals". Such judgments include, for example: "The domestic cat has four legs, two eyes, two ears, and guts in its belly", or "The Texas bluebonnet harbors nitrogen-fixing bacteria in certain nodes on its roots" (Thompson 2008, 64–65). Thompson is particularly interested in the logical form of such judgments, which is why, drawing on Frege, he endeavors to demonstrate that concepts such as "life" or "organism" are in fact logical forms and not, for example, concepts that denote the possible contents of more general logical forms:

> "It is because in the end we will have to do with a special form of judgment, a distinct mode of joining subject and predicate in thought or speech, that I am emboldened to say that the vital categories are logical categories." (Thompson 2008, 48)

What is important about Thompson's insights for my question is the fact that the judgments in which the forms of life are expressed are, on the one hand, normative judgments (in the very broad sense of criteria by which something is measured): A domesticated cat with only three legs is not an individual belonging to a species *other* than the species "domesticated cat"; rather, it is simply a defective specimen of the species "domesticated cat". It is this normativity that allows us to speak of "*the* domesticated cat" or "*the* bobcat" without intending thereby a reference to an inductive universality in the sense of "all (or most) domesticated cats", etc. The universality of the judgments relating to life forms is thus a *normative*, and not a descriptive, universality. On the other hand, they are judgments that always possess a diachronic character (therefore, a "natural-historical" one). The forms of life of a species can only be adequately described when the complete course of the life of an individual, typical to the species, is taken into account, including the characteristics of its propagation and its genesis.

I should be clear by now that, in Thompson's view (even if he sometimes furnishes it with a small question mark), the different *life forms* constitute the different *species*. Consequently, individuals belong to *a* species when they share *a* particular form of life. If affiliation to a species has any ethical meaning at all, it can only be assigned normative relevance if the concept of species is a concept of *life forms* that are different to each other. This is so, because forms of life in the Neo-Aristotelian sense have an essentially *teleological* character. That different species are characterized by different forms of life (in Thompson's sense) is what makes affiliation to species normatively relevant – precisely because each form of life exhibits a specific teleology. Moreover, it seems clear to me that the different *forms of life* are nothing other than different kinds and ways of realizing the ends of self-organization, self-preservation and propagation (i. e., self-reproduction and the reproduction of the species), and finally (in the case of higher animals), social life. These differing kinds and ways of the realization of ends are the very reason for different morphological properties. Mayr's population-genetic definition of species exploits, in turn, the important aspect of the reproduction of species itself, without which one can speak only of individuals, but not of species.

All these considerations are significant for the problem of transgenic organisms insofar as they permit the formulation of criteria whereby a transgenic organism may be considered to be a life form that transgresses the species boundary, that is, when it may be considered to be a real *cross-species* life form. At the same time, it permits the formulation of criteria which enables us to speak of what emerges as an *new* species. That is, it enables us to speak of a species that would no longer be that of those beings from whom the (now recombined) genes originated. Thompson's example of a three-legged cat (Thompson 2008) allows us to speak of this without having already to speak of genome editing. Initially, the cat is simply a defective specimen of the species "domesticated cat". However, if on the basis of its three-leggedness the cat were to come up with new strategies for self-preservation, preservation of its species and its social life, and if the three-leggedness, as well as the strategies based upon it could be consistently passed on to its descendants, then a *new species* will have emerged: That is to say, a group of beings determined by their own specific teleology and normativity, which defines them as beings. Expressed in the language of evolutionary biology, this new kind would inhabit a particular, hitherto uninhabited ecological niche.

The results of these findings can be carried out, without much ado, to transgenic organisms. Let us take as an example the manufactured being that has existed for more than two decades, the being which, through the insertion of a fluorescence gene, can under certain conditions generate light, namely the (in)famous GloFish. So long as this "edited-in" property of genetic fluorescent does not result in this fish developing a *new form of life*, by means of which it could be differentiated from its predecessors that remain unchanged by genetic technology, then we are not talking about a new species. *Per contra*, if on the basis of a property "edited-in" by gene technology a new form of life were to develop, and if this new form of life were to be passed on to its descendants, then we would be truly speaking of a new species, or at least a new subspecies.

6 The Normative Status of "Cross-Species Beings" and the Implications for the Argument from Species

It should be clear, in light of what has been said thus far, that I think that the majority of so-called "composite beings between man and animal", about which (unfortunately, often serious) newspapers are happy to write ominous headlines, are in fact no inter-species creatures at all. Think of the field of research, mentioned at the beginning of this chapter, which endeavors to cultivate "human" (from a genetic point of view) organs in pigs. As far as we know, the form of life of these pigs in no way differs from the form of life of other pigs. The fact that, for instance, the pancreas of such a pig exhibits a human rather than a porcine genome does not, as far as can be judged at present, have ramifications for the phenotype, let alone for the *form of life* constitutive of the species. Accordingly, we are not really speaking about "composite beings" in the sense of a composite of two different species; rather, we are clearly speaking about a creature of the species "domesticated pig", even if its pancreas is genetically human. This assessment is valid, and indeed all the more valid, regarding so-called "cytoplasmic hybrids" (cybrids). For instance, statements of the order that a cybrid is "up to 99% frog and 1% human being" do not make much sense. In as far as the one-percent amount of "human genes" in no way affects the phenotype, let alone the form of life of the creature concerned, we are simply dealing with nothing other than a frog. If the theory of the forms of life sketched above is correct, then it is completely impossible to quantify species affiliation in terms of percentage, for species affiliation is determined by differing forms of life and not by quantities of genes. This implies three possibilities for understanding the forms of life, and accordingly, species classification of transgenic beings:

1. A transgenic being whose genome originates from creatures of different species belongs to *one* of the species from which it originates. This pertains when the genetic component of the other species does not affect the phenotype, or rather, impacts it only in such a way that the phenotype does not generate a new form of life. As is evident, such variants present a problem neither for the question of the normative status of the creature concerned nor for the question of the validity of the bioethical argument from species – simply because no transgression of the species boundary here takes place.
2. A *new* species is formed when a *new* form of life emerges which is consistently passed on. In this case, too, no problem for the validity of the ethical argument from species is presented. With regard to the problem of the normative status of the creature of such a new species, it remains to be questioned whether the forms of life of the new species exhibit personality, or rather, subjectivity. In this case, they could be considered subjects of human rights, regardless of the species from which they are derived.
3. It represents a "defective" specimen of one or even several of the species from which it originates. The "defect" may lie in a deficient aptitude for self-organization (i. e., the bringing forth of itself as an organism in time), self-preservation, reproduction, or participation in the typical social life of a species, or, indeed, in a combination of several factors. If the form of life of one of the species from which it originates is characterized by personality, it can ultimately amount to a genetically induced impairment to developing the crucial characteristics of personality. This final subvariant is, as mentioned above, the critical one.

Let us begin by considering the implications of possibilities (1) and (2): Although, to my mind, cases (1) and (2) raise no basic *ontological* questions, this might not be the case regarding serious *epistemic* problems. In certain circumstances, it might prove difficult to recognize forms of personality that differ from those which are familiar to us (this problem already exists in the cases of dolphins and higher primates). It is, therefore, theoretically conceivable that genetically modified domesticated pig hybrids could produce characteristics of personality. This might happen if, contrary to expectations, human iPS cells (introduced in order to cultivate genetically human organs) were to assist in the formation of the structure of the brain and nervous system. In this case, it might happen that the hybridized domesticated pig would, as such, no longer belong to the species "domesticated pig". Rather, it might form a new species, the form of life of which would be characterized, as in human beings, by personhood. However, it is conceivable that this would not be conspicuous to us for a long time. To me, this problem of epistemic uncertainty already appears to be a good argument against the manufacture of "composite beings" between the animal and the human. The danger here is obviously that, in certain circumstances, we fail to recognize beings that have a claim to recognition as subjects of human rights, because we do not perceive them as persons at all. At this time, this epistemic problem appears to me to be the most striking argument against the manufacture of transgenic beings with certain (from the genetic point of view) "human" components.

Something similar pertains to possibility (3): If one were to undertake the genome editing of a specimen of the species "human", on the basis of which this being would with great probability not possess, or would no longer possess the capacity to develop personality in the course of its life, such a being would not, as a result, be a specimen of *another* species. On the contrary, it would be a "defective" specimen of the species "human". It is not only that considerable epistemic uncertainties would emerge from the start, but also, and much more, that the central argument would be that such an intrusion amounts to nothing other than a massive bodily injury, with the consequence that the victim loses his personality.

7 Conclusion

In conclusion of these considerations, four points can be maintained:

1. The manufacture of transgenic beings, that is, a "mixing" of the genes of different species on the genetic level, does not lead to the emergence of an interspecies "composite being". Transgenic beings, too, always belong to a species, be it to an existing one, or to one newly constituted.
2. The validity of bioethical arguments from species, in both of their strong variants (the Neo-Aristotelian and the transcendental-pragmatic), is not called into question by the possibility of the manufacture of transgenic beings with (genetically speaking) a "human" component.
3. The creation of transgenic beings with (genetically speaking) a human component should take place only where epistemic uncertainty, about whether the beings in question can exhibit or cultivate personality, can be excluded.
4. The deliberate manufacture of "defective" specimens of the species "human" is to be rejected, because it amounts to massive bodily injury of the being in question.

References

Fichte, Johann G. 2000. *Foundations of Natural Right According to the Principles of the Wissenschaftslehre*. Neuhouser, Frederick (ed.). Cambridge/New York: Cambridge University Press.

Hegel, Georg W. F. 1977. *Phenomenology of Spirit*, Inwood, Michael (ed.), Oxford: Oxford University Press.

Mayr, Ernst. 1963. *Animal Species and Evolution*. Cambridge: Harvard University Press.

Müller-Jung, Joachim. 2017. Der Mensch im Schwein. *Frankfurter Allgemeine Zeitung*, January 27.

Prinz, Wolfgang. 2012. *Open Minds. The Social Making of Agency and Intentionality*. Cambridge: MIT Press.

Richards, Richard A. 2010. *The Species Problem: A Philosophical Analysis*. Cambridge/New York: Cambridge University Press.

Schweidler, Walter. 2008. Ethische Neutralität und weltanschauliche Souveränität des Staates. In *Begründung von Menschenwürde und Menschenrechten*, Härle, Wilfried, and Bernhard Vogel (eds.), 169–187. Freiburg: Alber.

Spaemann, Robert. 2000. *Happiness and Benevolence*. Edinburgh: T&T Clark.

Spaemann, Robert. 2006. *Persons. The Difference between 'Someone' and 'Something'*. New York: Oxford University Press.

Thompson, Michael. 2008. *Life and Action: Elementary Structures of Practice and Practical Thought*. Cambridge: Harvard University Press.

Germline Modifications as a Severe Intervention into Human Nature

Nadia Primc

Keywords

Genome Editing, Germline Manipulation, Human Nature, Philosophy of Science, Technoscience, Biofact

Abstract

Manipulation of the human germline is sometimes criticised as a severe intervention into human nature. In order to assess the tenability of this claim, different uses of the term "human nature" and distinct scientific and clinical contexts must be differentiated. The first live birth of an edited human child will probably take the form of an unproven intervention and not a clinical trial. As an unproven intervention, germline manipulation must be regarded as a severe intervention into the nature of a prospective newborn. Nevertheless, once the technique of germline manipulation can be regarded as a safe therapeutic option, and if its use is restricted to the prevention of severe genetic diseases, it can no longer be regarded as a particularly severe intervention into human nature. However, such a use would make human nature the object of technoscience and the edited human beings would have to be regarded as biofacts, i. e., a form of incarnated and living technological design. This introduces a new quality into the notion of "human nature", which can be regarded as ethically acceptable, as long as the intervention is restricted to the prevention of severe genetic diseases.

M. Braun et al. (Eds.), *Between Moral Hazard and Legal Uncertainty*, Technikzukünfte, Wissenschaft und Gesellschaft / Futures of Technology, Science and Society, https://doi.org/10.1007/978-3-658-22660-2_7

1 Introduction

Systematic manipulation of the human germline is described by some critics as a severe intervention into human nature. In order to assess the tenability of this claim, different uses of the term "human nature" must be differentiated (Chapter 2).

The same applies to the various scientific and clinical contexts in which manipulation of the human germline supposedly occurs. In the ethical debate about the permissibility of genome editing, it is commonly argued that the consequences of manipulating the human germline are much harder to predict, and potentially affect a greater number of human beings, than somatic gene therapies (The National Academies 2017, 3). This raises the question of whether and how the technique of human germline manipulation can make the transition from a theoretical concept to a safe therapeutic option. This issue and the question of whether this transition can be regarded as a severe intervention into human nature will be discussed in Chapter 3.

The claim that manipulation of the human germline represents a severe intervention into human nature is frequently justified by interpreting the manipulation as an impermissible instrumentalisation, or an infringement of the individual's right to autonomy. Some aspects of these deontologically inspired claims will be discussed in Chapter 4. Chapter 5 addresses the closely related objection that genome editing is making human nature the object of technoscience.

All the considerations in this paper are restricted to the use of systematic human germline manipulation to minimize the risk of severe genetic disorders. Although it is not possible to provide a precise definition, or definite list, of what should be regarded as a severe genetic disease, it will be postulated, for the sake of the argument, that it is possible to agree on a set of candidates that should be regarded as severe genetic diseases. For reasons of brevity, this will be referred to throughout this paper as a therapeutic use of human germline manipulation.

2 The Term "Human Nature"

An obvious difficulty in giving an account of the different uses of the term "human nature" is the fact that the definition of the term "human nature" is the subject of ongoing philosophical debate.[1] Furthermore, there is no consensus as to whether

1 Some scholars see human nature as an integral part of nature in a broad sense (Heyd 2005, 52–54). The term "human nature" is sometimes used to designate a biological

human nature has an inherent value or merely a derivative normative status (Bayertz 2003, 140–142). All of the philosophical attempts to define human nature face a number of problems too great to discuss here. It is likewise impossible to provide an elaborate account of the concept of human nature here. For the purposes of the present inquiry it will suffice to differentiate between two basic uses of the term "human nature".[2]

First of all, the term "human nature" can be used in a strictly empirical sense, i.e., to designate the totality of the physical and psychological traits of human beings. These traits can be the object of an empirical description, or a scientific investigation, in the different branches of biology, psychology, medicine, cultural anthropology, political science, etc.[3] In such an empirical account, there are no empirical characteristics that can be regarded as part of a substantial or unchanging essence of human beings, since, in a post-Darwinian age, they must be considered the product of ongoing natural selection (Roughley 2005, 137–139). It should be clear that any medical intervention represents an intervention into human nature in this empirical sense, inasmuch as it attempts to positively influence certain empirical characteristics of the patient (Heyd 2005, 63). In order to counter the technique of germline modification it is therefore necessary to demonstrate that it represents a particularly severe intervention into human nature and is substantially different to other biomedical interventions.

On the other hand, the term "human nature" can also be used in a non-empirical and therefore more metaphysical sense,[4] particularly if the aim is to provide evidence of a specific or universal human essence. In this case, the term "human

species. Sometimes it is used to describe those traits and characteristics that are not artificial, i.e., the result of deliberate human manipulation (Roughley 2005, 137–139).

2 Heyd appears to make a similar distinction (Heyd 2005, 70–71).

3 This does not mean that disciplines such as cultural anthropology or political science should restrict themselves to an empirical description of human beings. Nevertheless, if they were to move beyond such empirical investigation and theory, they would no longer be concerned with human nature in an empirical sense.

4 The term "metaphysical" has been given a wide range of meanings within the history of philosophy. Here, it is used in a negative and somewhat Popperian sense, i.e., to designate a claim or description that cannot be easily falsified by empirical observation (Popper 2005, 10–12). The postulation of a human essence must be regarded as a metaphysical claim, since apparent empirical counter-examples (e.g., anencephalic newborns, patients with severe neurological damages), being identified as human but lacking the relevant capacities (autonomy, reason, etc.), are generally not accepted by the defenders of the idea of a common human essence as a falsification of their notion. Rather, such counter-examples are viewed as exceptions, or special cases, that must somehow be explained within their own account of human nature.

nature" designates a set of characteristics and capabilities that substantially distinguishes human beings from other living beings and cannot be fully captured in empirical terms (Bayertz 2003, 137).[5] This of course raises the difficult philosophical question of whether it is possible to identify any unique characteristics, for the purpose of drawing a sharp line between humans and other living beings. The most promising candidates are probably the capacities for reason, autonomy, and morality, which are all closely linked to each other and to the consideration of human beings as persons, or as beings with a special moral status and moral capacity.[6] This non-empirical account is closely connected to the conception that human nature is essentially artificial, i.e., that an important goal or the telos of human life is to protect, actively develop and cultivate these capacities in oneself and others by means of education, technology, science, etc. (Heyd 2005).

Of course, any biomedical intervention that negatively interferes with any of these capacities must *prima facie* be regarded as ethically questionable (Heyd 2005, 64–65). However, it should also be clear that any biomedical intervention that can successfully prevent severe genetic diseases in an individual has the potential to increase the prospective newborn's chance of a healthier life, and hence, its ability to exercise and develop its capacity for reason, autonomy, and morality (Heyd 2005, 57). This is why health is sometimes referred to as an enabling good and given a special normative value.

3 First-in-Man Use of an Unproven Intervention

An important question to be asked with regard to the possibility of therapeutic human germline manipulation is how the transition from a theoretical concept to a relatively safe medical option can be achieved in clinical practice.

It will of course be necessary to conduct further experiments on animals in order to acquire greater knowledge and control of off-target effects, mosaicism and the long-term consequences of germline manipulation (The National Academies 2017).

5 Although this does not rule out the possibility of providing a non-exhaustive description of these characteristics in empirical terms, any account that claims to fully capture the capacity for reason, autonomy, morality, etc. in empirical terms must by definition be subsumed under the aforementioned empirical accounts of human nature.

6 Other possible candidates for such a *differentia specifica* include language, self-consciousness and the contention that human beings are political animals. These conceptions will not be discussed in detail here as the following considerations can invariably also be applied to the various accounts.

Given the limited transferability of the results from animal research to human development, it is often argued that further experiments should be conducted on human embryos (The National Academies 2017).[7] In order to gain a greater number of human embryos for this type of research, and in order to ensure the subsequent live birth of at least one child bearing the intended genetic traits in each case in clinical practice, some scientists urge the need to combine genome editing with in vitro gametogenesis, and specifically, the need to generate eggs from induced or adult stem cells (Hikabe et al. 2016). Further investigation is necessary to improve the evaluation of the short- and long-term consequences of in vitro gametogenesis (The National Academies 2017, 78–79; 243–244).

Determining the point at which results in live animals and human embryos can be deemed good enough to risk the first live birth of an edited human being is a very challenging ethical decision. Even the advocates of germline manipulation agree that "it would be highly unethical to bring modified human embryos to term unless we were very confident that the technique could be used safely. The risk would simply not be justified by any potential benefits" (Savulescu et al. 2015, 477).

Although proof of principle is possible within preclinical research, the first use in humans must be regarded as a risky experimentation, especially because human beings live much longer and are exposed to a greater variety and amount of environmental impacts than any in vitro or in vivo model used within preclinical research.

Those who counter these concerns sometimes observe that similar scepticism about potential long-term effects once existed with regard to in vitro fertilization (IVF) (specifically in the case of the birth of Louise Brown, the world's first "test tube baby", in 1978) and certain techniques of preimplantation genetic diagnostics (PGD), e.g., the blastomere biopsy, which involves a considerable reduction of the embryo, as it removes one or two cells from the eight-cell embryo. As the advocates of the human germline manipulation like to point out, both techniques proved to be safe, even though they were once criticized as being highly unsafe (Bernard 2014, 373–375, Savulescu et al. 2015, 477). Nowadays, IVF and PGD are perceived as an almost equivalent option to natural reproduction (Roughley 2005, 147).

Whatever the force and scope of these historical counter-examples, they draw our attention to the fact, that as in the case of Louise Brown, the first live birth of an edited human child will probably take the form of an unproven intervention

7 Some preclinical studies that use CRISPR to induce germline modifications in human embryos have already been published (Liang et al. 2015, Kang et al. 2016, Tang et al. 2017, Ma et al. 2017) and further publications are anticipated.

and not of a clinical trial.[8] According to Article 37 of the Declaration of Helsinki, an unproven intervention should be restricted to cases "where proven interventions do not exist or other known interventions have been ineffective" and "if in the physician's judgement it offers hope of saving life, re-establishing health or alleviating suffering" (World Medical Association 2013).

These requirements do not appear to apply to the case of therapeutic germline manipulations. First, IVF and PGD, as well as egg and sperm donation have already been proved to be sufficiently safe and effective options. Second, there is no pre-existing individual whose suffering is alleviated or life is saved with the help of germline manipulation. Each embryo must be deliberately and artificially generated by means of IVF with the goal to subsequently apply the technique of germline manipulation. The transition to a safe clinical option involves the challenging ethical situation of a prospective newborn having to undergo all the risks of an unproven intervention, just to alleviate the distress caused to its parent by their intense wish to have a genetically related child. This raises certain questions about the ethics of parenthood (Haker 2013, Habermas 2013, 29–30), its relation to the concept of reproductive autonomy, and the extent to which clinical practices and clinical research should be bound to these considerations. Despite the possibilities the technology of germline manipulation opens up for the advancement of biomedical science and our understanding of the role of genetic information within human development, as an unproven intervention, manipulation of the human germline must surely be regarded as a severe intervention into the nature of prospective newborns.

4 The Manipulation of the Human Germline as a Severe Intervention into Human Nature

In contrast to the preceding chapter, the following considerations are restricted to the hypothetical case where the manipulation of the human germline can be regarded as a sufficiently safe clinical option, i. e., where the associated risks are minimal and the birth of a healthy newborn can be expected to be the "normal" case.

As mentioned in the introduction, the claim that manipulation of the human germline represents a particularly severe intervention into human nature is com-

8 I therefore strongly disagree with the National Academies, which seems to assume that the implementation of germline manipulations in clinical practice will take the form of closely monitored clinical trials (The National Academies 2017, 45–47).

monly justified by interpreting it as an infringement of the individual's right to autonomy, or an impermissible instrumentalisation of human beings (The National Academies 2017, Habermas 2013).[9] The crux of this argument is the potential severity of the impact[10], on the lifeworld of a human being, caused by a manipulation of the human germline. As noted above, systematic germline manipulation must be identified as a particularly severe intervention, since every medical intervention, by definition, represents an intervention into human nature in the empirical sense of the term.

The primary aim of genome editing is to cure genetic diseases and allow couples with an unfavourable genetic disposition to have children who are genetically related to both of them. It is often pointed out that the latter (at least in some countries) already have some other reproductive medicine techniques at their disposal, e.g., the option of sperm or egg donation (Savulescu 2015, 476; The National Academies 2017, 113–115). Such currently practised reproductive medicine techniques are still criticised by donor-conceived persons for several reasons, including the effects they could have on the child's psychological development and its everyday sense of identity.[11] In Germany, *Spenderkinder*, an organisation of persons who are the offspring of an anonymous sperm or egg donation, are committed to giving a voice to such persons. Although some of these persons claim to have a close relationship with their social parents, the fact that an unknown donor has played a role in their genetic origin represents a particular challenge to their everyday sense of personal identity and their personal history (Spenderkinder 2006). However, this does not mean the unknown donor needs to play an important role in the child's development, or that this impact needs to be grasped in psychopathological terms (Meier-Credner 2016, 156). As *Spenderkinder* and the persons concerned stress, these issues may only become relevant at a later stage in life, e.g., when donor-conceived persons

9 The same applies to the claim that manipulation of the human germline represents a violation of human dignity. This claim must also be explained and is often justified by qualifying the manipulation of the human germline as an infringement of the right to autonomy, or an impermissible instrumentalisation of human beings.

10 It seems inconsistent and hardly tenable to define something as a severe intervention into human nature even though there are no potential effects on individuals' lifeworlds.

11 Several studies and systematic reviews suggest that donor-conceived children do not differ in any significant way in their psychological development compared to naturally conceived children (Illioi and Golombok 2014). Nevertheless, most of these studies are subject to a number of shortcomings. One point of criticism is that the surveys "tend to have small sample sizes that are possibly biased to include people who are functioning well" (Illioi and Golombok 2014, 94). More important is the fact that only a low percentage of the families included in those studies have informed their children about their mode of conception (Illioi and Golombok 2014, 93).

begin to plan families of their own. Just because such people feel a need to come to terms with the fact that they are the product of an anonymous sperm or egg donation does not mean that they approve of every aspect of their personal genesis (Meier-Credner 2016, 156). Being the offspring of an anonymous donor definitely has a potential effect on the lifeworld of donor-conceived persons. However, the scope and severity of this impact may vary substantially between individuals and cannot be reliably predicted.

The option of genetic engineering could avoid the challenges of unknown genetic heritage, since such persons would not be forced to inquire into the identity of a third party that forms a substantial part of their genetic make-up. The doctor or biotechnologist performing the genome editing in the laboratory will certainly not have the same impact on the child's personal history and everyday sense of identity as some unknown donor, who may, meanwhile, have a family and children of his or her own.

A similar line of reasoning can be applied to the claim that manipulation of the human germline represents an instrumentalisation of human beings. One must either regard every IVF that is combined with PGD as an impermissible instrumentalisation (a position that is hard to defend in a pluralistic society), or one must conclude that the addition of therapeutic genome editing to currently available technologies does not constitute an additional aspect of instrumentalisation.

If systematic manipulation of the human germline can be regarded as a sufficiently safe clinical option, and if it is used only to prevent severe genetic diseases, it cannot really be regarded as a threat to the basic capacity for autonomy, reason, morality, etc., or to any other faculty or characteristic that is considered an important part of our human essence.[12]

5 Human Nature as an Object of Technoscience

The term "technoscience", as it is used here, refers to scientific research programmes driven mainly by available technological options, and in which scientific knowledge is produced primarily by implementing these technologies. This implies that the pace and direction of scientific progress is determined by the pace and direction

12 This reflects Habermas' position. He argues that individuals will, later in life, be likely to approve of prenatal measures taken to protect them from severe diseases (Habermas 2013, 109).

of technological development itself, and less by factors that are external to technological progress, e.g., the medical needs of patients.

The term "technoscience" emphasises that the influence of technology within science is growing because, to a large extent, knowledge is generated through the use and implementation of technology. The term also draws our attention to the fact that scientific knowledge is objectified to a considerable degree in technological interventions. Within technoscience, it is no longer possible to draw a sharp line between nature and the technological interventions themselves (Nordmann 2014, 31–33). One famous example is the so-called oncomouse, which is created by introducing an oncogene that is designed to trigger the growth of tumours. The term "biofact" was coined for this sort of genetically designed biological entities (Karafyllis 2003). The initial goal of these experiments is to demonstrate that it is possible to gain a certain amount of control over natural phenomena.[13] If successful, the results speak for themselves and a transition to different branches of applied research can be envisaged. Within technoscience it is no longer possible to differentiate in a traditional way the characteristics and nature of organisms from the technical interventions and design required to create and to study them. Their characteristics are equally part of their technologically designed nature and part of the technical intervention itself (Nordmann 2014, 35).[14]

Under the regime of technoscience, scientific research consists mainly of attempting to make the available technologies fit existing social or medical demands (Nordmann 2014, 33–34). Its goal is to provide a technological fix for the latter,

13 The use and design of animal models is generally justified by their supposed benefits for the development of medical treatments. Nevertheless, the initial incentive for the design of animal models lies within the general technological possibility of genetically modifying living organisms. This technology has been developed to a large extent independently from the different areas of implementation (industry, agriculture, food production, pharmaceutics, medicine etc.). After proof of principle, the validity of new technologies for different branches of applied research can be investigated. This second step opens up the possibility of further refinement and development of these technologies.

14 The modern sciences have, of course, always used technologies to intervene in natural processes, as this is a hallmark of scientific experimentation. Nevertheless, in the traditional image of modern science, experimental design is used and understood as a trigger for a controlled natural process, within which nature is forced to disclose some of its basic characteristics and dispositions. Nature reveals itself within the experimental set-up. The technical interference can (at least to a certain extent) be differentiated from the dispositions and characteristics that lie within nature and which can be made the object of scientific theories. Within technoscience, technological interventions become an inseparable part of the nature of the studied objects themselves (Nordmann 2014, 30).

since this gives the researchers the opportunity to further refine the technology and scientific knowledge.

How does this apply to the topic of genome editing? In order to refine the technique of genome editing, and the knowledge of the role of genetics within human development, basic research needs a medical application that more easily justifies, in both moral and financial terms, the preclinical development and refinement of genome editing in humans as well as its clinical implementation. Whereas these techniques were initially adopted by human medicine to offer patients with severe and untreatable genetic diseases the option of a somatic gene therapy, the idea of extending these technologies to include the manipulation of the human germline developed very quickly.[15]

Human reproduction and human nature, in the empirical sense discussed above, are thus made the object of technoscience, which gains its knowledge through the development and implementation of a specific technique. Hence, we are confronted with the paradox that knowledge of the role of genetics in human development, and the possible short- and long-term effects of germline editing, can only be gained through preclinical and clinical implementation of these techniques in human beings, even though this knowledge is often required as an ethical precondition for the implementation of these techniques in humans. The required knowledge can only be gained and objectified in the edited human beings, who must then be regarded as biofacts, i.e., a form of incarnated and living technological design. Being the object of technological design introduces a new quality into the notion of "human nature" and establishes a new form of relationship between human beings, namely between a technological designer and a designed human being.[16] This does not imply that this new quality of human nature must be regarded as ethically objectionable – as long as the intervention is restricted to the prevention of severe genetic diseases (see Chapter 4).

---

15 This point is supported by the fact that all the published studies involving manipulation of the human germline are related to specific genetic diseases (Liang et al. 2015, Kang et al. 2016, Tang et al. 2017, Ma et al. 2017).

16 The same question, namely whether the patients must be regarded as biofacts, could be asked in the case of certain forms of somatic gene therapies, especially if the induced changes are intended to be permanent. Nevertheless, the differences between somatic gene therapy and germline manipulation are relevant in terms of ethics. For instance, informed consent can be obtained by participants in clinical trials that focus on somatic gene therapy. Of course, this is not possible for children or young adolescents. However, in contrast to germline manipulation (see Chapter 3), somatic gene therapy is generally used to alleviate the suffering or save the life of an existing individual.

6 Conclusion and Outlook

The present inquiry has revolved around the question of whether manipulation of the human germline can be regarded as a severe intervention into human nature. The answer to this question depends on the clinical and scientific context in which such manipulation supposedly occurs. As an unproven intervention, germline manipulation must be regarded as a severe intervention into the nature of a prospective newborn. Nevertheless, this does not hold for the hypothetical case that germline editing can be regarded as a safe therapeutic option if its use is restricted to the prevention of severe genetic diseases. Furthermore, it has been argued that manipulation of the human germline would make human nature the object of technological design, which introduces a new quality into the notion of "human nature". This newly established relationship between a designer and a designed, edited being can be regarded as ethically defensible, as long as the intervention is restricted to strictly therapeutic use.

But what about a possible extension of germline editing beyond such strictly therapeutic use? In order to expand our knowledge of the role and interplay of our genetic information, the technique must be extended to other sorts of conditions, and to induce certain desirable characteristics in human beings, e.g., a stronger immune system or resistance to certain diseases and afflictions. In the long run, the advancement of scientific knowledge in the field of genetics demands that germline manipulation be extended from largely therapeutic applications to preventative applications, as well as to the cause of genetic enhancement.

This could introduce a new form of social injustice between those who can afford these techniques of genetic enhancement and those who cannot (Heyd 2005, 68; The National Academies 2017, 127–128), which could also pose a risk to moral cohesion within our societies (Habermas 2013, 115). Furthermore, the possibility of genetic enhancement demands a reassessment of the aforementioned arguments concerning the potential infringement of the individual's right to autonomy and an open future (Heyd 2005, 68, Habermas 2013, 109).

A common refutation of these claims, regarding the technology-driven development of this field of research, cites the potential for political and legal regulation of biomedical technologies. It is argued that technological development is not an untameable process, but one that societies can actively shape. Nevertheless, this argument underestimates the globalisation of health markets and the issue of medical tourism, which is currently exemplified by reproductive tourism. "The result could be a 'race to the bottom' that would encourage laxer standards in nations seeking revenues from medical tourism, as has happened with both stem cell therapy and mitochondrial replacement techniques" (The National Academies 2017, 135).

References

Bayertz, Kurt. 2003. Human Nature: How Normative Might It Be? *Journal of Medicine and Philosophy* 28: 131–150.

Bernard, Andreas. 2014. *Kinder machen. Samenspender, Leihmütter, Künstliche Befruchtung. Neue Reproduktionstechnologien und die Ordnung der Familie*. Frankfurt am Main: Fischer.

Habermas, Jürgen. 2013. *Die Zukunft der menschlichen Natur. Auf dem Weg zu einer liberalen Eugenik?* Frankfurt am Main: Suhrkamp.

Haker, Hille. 2013. Eine Ethik der Elternschaft. In *Kinderwunsch und Reproduktionsmedizin. Ethische Herausforderungen der technisierten Fortpflanzung*, Maio, Giovanni, et al. (eds.), 269–292. München: Karl Alber.

Heyd, David. 2005. Die menschliche Natur: ein Oxymoron? In *Die menschliche Natur. Welchen und wieviel Wert hat sie?* Bayertz, Kurt (ed.), 52–72. Paderborn: Mentis.

Hikabe, Orie, et al. 2016. Reconstitution in Vitro of the Entire Cycle of the Mouse Female Germ Line. *Nature* 539: 299–303.

Illioi, Elena Christiana, and Susan Golombok. 2015. Psychological Adjustment in Adolescents conceived by Assisted Reproduction Techniques: a systematic Review. *Human Reproduction Update* 21: 84–96.

Kang, Xiangjin, et al. 2016. Introducing Precise Genetic Modifications into Human 3PN Embryos by CRISPR/Cas-mediated Genome Editing. *Journal of Assisted Reproduction and Genetics* 33: 581–588.

Liang, Puping, et al. 2015. CRISPR/Cas9-mediated Gene Editing in Human Tripronuclear Zygotes. *Protein Cell* 6: 363–372.

Ma, Hong, et al. 2017. Correction of a Pathogenic Gene Mutation in Human Embryos. *Nature* 548: 413–419.

Meier-Credner, Anne. 2016. Was bedeutet Embryonenadoption aus der Perspektive der entstehenden Menschen? *Zeitschrift für medizinische Ethik* 62: 151–164.

Nordmann, Alfred. 2014. Das Zeitalter der Technowissenschaften. In *Strukturwandel der Wissenschaft. Positionen zum Epochenbruch*, Nordmann, Alfred, et al. (eds.), 24–38. Weilerwist: Velbrück Wissenschaft.

Popper, Karl Raimund. 2005. *Logik der Forschung*. Tübingen: Mohr Siebeck.

Spenderkinder. 2006. Meinungen und Geschichten. http://www.spenderkinder.de/ueberuns/meinungenundgeschichten. Accessed 08 March 2018.

Roughley, Neil. 2005. Was heißt „menschliche Natur"? Begriffliche Differenzierungen und normative Ansatzpunkte. In *Die menschliche Natur. Welchen und wieviel Wert hat sie?* Bayertz, Kurt (ed.), 133–156. Paderborn: Mentis.

Savulescu, Julian, et al. 2015. The Moral Imperative to Continue Gene Editing Research on Human Embryos. *Protein & Cell* 6: 476–479.

Tang, Lichun, et al. 2017. CRISPR/Cas9-mediated Gene Editing in Human Zygotes Using Cas9 Protein. *Molecular Genetics and Genomics* 292: 525–533.

The National Academies of Sciences, Engineering, and Medicine. 2017. *Human Genome Editing. Science, Ethics, and Governance*. Washington DC: The National Academies Press.

World Medical Association. 2013. *Declaration of Helsinki – Ethical Principles for Medical Research Involving Human Subjects*. https://www.wma.net/policies-post/wma-declaration-of-helsinki-ethical-principles-for-medical-research-involving-human-subjects/. Accessed 28 October 2017.

Is Genome Editing Unnatural?

Nature in Bioethics, the Politics of Bioethics, and the Political Dimension of Nature[1]

Eva Odzuck

Keywords

Genome Editing, Naturalness, Bioethics, Political Dimension of Nature, Political Theory of the Body

Abstract

The concept of *nature* and *naturalness* plays an important role in academic and public bioethical discussions. Given the obvious weakness of arguments that are based on a genetic and/or totalizing view of *nature*, the immense and repeated efforts to criticize them demand explanation. The repeated critique of such weak arguments and the tendency to avoid the category of nature can be explained (at least in part) by political motivations. While an unjustified appeal to *nature*, or a description of genome editing as *unnatural*, is of course problematic, we cannot develop the contours of a responsible genome policy without the dimension of *nature*. I work with the hypothesis that our relation to *nature* is in itself deeply normative and political, and conclude that a responsible biopolicy requires us to bring the *question* of *nature* back into the center of our political philosophy *and* our liberal democracies.

1 I want to thank the members of the winter school "Between moral hazard and legal uncertainty – Ethical, legal and societal challenges of gene editing" (February 13th-17th, Erlangen-Nuremberg), as well as Christoph Herrler, Clemens Kauffmann, Alexander Necker, Christian Seidel and the editors of this volume for their helpful comments on my talk and/or manuscript.

M. Braun et al. (Eds.), *Between Moral Hazard and Legal Uncertainty*, Technikzukünfte, Wissenschaft und Gesellschaft / Futures of Technology, Science and Society, https://doi.org/10.1007/978-3-658-22660-2_8

1 Introduction

The rapid scientific developments with respect to technical possibilities of genome editing (for example, the CRISPR-Cas9 method) have changed the context of a debate that has hitherto remained rather hypothetical: the debate about the modification of and intervention into the human germline. Technological development, the recent results of international research groups,[2] economic considerations, and the hope that one day what are sometimes called *genetic scissors* could help to heal genetic diseases, all intensify the urgency for liberal democracies to think about the normative desirability and legitimacy of germline intervention in the human embryo. Early in the debate there was a lot of caution and reluctance. In addition to the reservations voiced by international bioethics committees, civil society organisations and religious representatives, there were also leading research groups who expressed the necessity of intensive reflection, advocating moratoria with articles entitled, "Don't edit the human germline" (Lanphier et al. 2015), or "A prudent path forward for genomic engineering and germline gene modificatio*n*" (Baltimore et al. 2015). In its latest *ad hoc* recommendation, the German Ethics Council warns that this caution has vanished in light of recommendations published by the *US National Academies of Science, Engineering and Medicine*, which "reveals a [...] shift in the evaluation of ethical accountability: [...] Apparently, speculations now concentrate less on whether but rather only on when the first human genetically modified by genome editing will be born" (German Ethics Council 2017, 3).

The speed and dynamic of technological successes and scientific developments seem to have silenced the calls for moratoria, the question *whether* seems to have been replaced by the question *when*, and fundamental considerations seem to have given way to a simple technical risk analysis. One factor possibly reinforcing this shift in the debate,[3] and the lack of fundamental perspectives, might be a recent trend in academia of being overcautious about *nature* as a category of philosophical and public debate. It is not only the transhumanist visionaries, but also an important contingent of mainstream Anglo-Saxon bioethicists and renowned ethic councils,

2 In August 2017, an international research group from Oregon published their research findings on the germline treatment of a dominant hereditary disorder. In 2015 and 2016, Chinese research groups had published their research findings about experiments with genome editing (the correction of a genetic predisposition for the blood disorder thalassemia and the generation of genetic resistance to HIV). For detailed information and references, see (German Ethics Council 2017, 2).

3 The chair of the German Ethics Council, Peter Dabrock, warns with good reason that there is not nearly enough debate, and that politicians and the democratic public should set the question of genome editing on their agenda (Dabrock 2017).

who point out conceptual problems with the term *nature*. They demand that these terms should not be used "without conveying the values or beliefs that underlie them", and warn that "effective communication on the ethics of science, technology, and medicine may be hindered, rather than helped, by appeals to naturalness" (Nuffield Council on Bioethics 2015, 107–108; for a similar claim compare Lustig 2009, 30.) For the influential bioethicist Savulescu, the case seems to be simple (cf. Savulescu 2009, 526): If neither god nor nature can be accepted as arguments in the public policy decision making of liberal democracies, there seems to be no reason whatsoever to be overcautious about the new technical possibilities. According to some transhumanists and self-declared bioliberals, the new times demand imitating Prometheus, the courageous use of the new biotechnologies for the amelioration of the human condition.[4]

But, as usual, the story is more complicated than the oversimplified dividing line drawn by some bioliberals between *bioconservatives* (appealing to *nature and intuitions)* and *bioliberals* (pursuing an analytic, scientific approach) suggests.[5] Taking as my premise that the category of *nature* is of utmost importance for political decisions about human life and the living human body, I argue for the necessity of bringing the *question* of *nature* and of the human relation to *nature* back into the center of our philosophical and normative discussions. I proceed in the following way:

In a first step, I present two different positions on the question whether *naturalness* sets limits to genome editing. While one side claims that *naturalness* is a problematic category in bioethical argumentation, being therefore also problematic with respect to genome editing, the other claims that there can be a necessary and reasonable use of arguments working with an idea of *naturalness* in bioethics and in debates about genome editing.

In a second step, I broaden the perspective and try to make sense of a widespread tendency, in current bioethical research (and political philosophy), to be reluctant about the category of *nature* in general and *human nature* in particular. I argue

4 Cf. Dworkin (1999): "Playing God is indeed playing with fire. But that is what we mortals have done ever since Prometheus, the patron saint of dangerous discovery. We play with fire and accept the consequences, because the alternative is an irresponsible cowardice in the face of the unknown."

5 Compare Birnbacher, who employs this dividing line, although he concedes that "this is an oversimplification" (Birnbacher 2014, esp. 82, 150). An oversimplified perspective can be found also in Clarke and Roache (2009). Ranisch (2013), is very nuanced in his treatment and offers helpful categories to differentiate between the two ideal types. Transhumanists prefer to divide the field into bioconservatives and *transhumanists* (Bostrom 2005, 202–204).

that this "case against nature" is often (also) motivated by political considerations, and in this sense, those who argue thus pursue a philosophy that is political in a problematic sense.

In a third step, I argue that the category of *nature* is indispensable for normative reasoning in general, and therefore, also for bioethics *and* political philosophy. A notion of *nature* that conceives *nature* as malleable, complex and consisting of different "selves" avoids essentialist oversimplifications and is not *per se* anti-democratic or illiberal. Rather, it calls for the spelling out, explanation and justification of the human relation to this malleable *nature* in general, and in particular, to the human body as that part of *nature* that is closest to the human person.

2 Nature, Bioethics and Political Philosophy

2.1 Does Naturalness Set Limits to Genome Editing? Two Different Positions

At the annual meeting of the German Ethics Council (entitled "Access to the human genome. New possibilities and their ethical evaluation") in Berlin in June 2016, two German philosophers took two different positions on the question whether *naturalness* sets limits to genome editing: Dieter Birnbacher, an analytic philosopher who has researched the role of *nature* and *naturalness* in bioethics (cf. Birnbacher 2014; 2006a, b; 1997), and theologian and philosopher Eberhard Schockenhoff, who developed an approach called "the ethics of life" (cf. Schockenhoff 2013). Although their positions were presented in short, oral statements, often suggesting a line of argumentation rather than fully developing it, I think it might be helpful to present their positions as a starting point for our question. For they can sensitize us to the fact that both positions rest on different normative premises and work with completely different notions of *nature*.

2.1.1 Dieter Birnbacher

Birnbacher argues that naturalness is a weak argument in many contexts, including that of genome editing.[6] He directs his argumentation against a very special

6 All citations from Birnbacher and Schockenhoff in this chapter (2.1) are from the *Simultanmitschrift* (notes taken down simultaneously) (German Ethics Council 2016, 70–79).

notion of *nature* that might provisionally be described as totalizing and genetic.[7] He disputes the normative relevance of a genetic notion of *nature*, citing Mill's umbrella argument: Since an umbrella is not part of (God's created) *nature*, but a human product, those who would take *nature* as their normative point of orientation would have to forbid the umbrella. He presupposes a totalizing view of *nature* in his second argument, according to which whoever tolerates interventions into the *nature* of plants or animals cannot, in the name of nature, consistently forbid interventions into the *nature* of humans, or the human germline. Afterwards, Birnbacher argues against positions that employ the category of *nature* to defend a restrictive position on germline editing. Based on a distinct theory of personhood, he argues along Parfit's lines, against Habermas, that an intervention into the germline of an unborn person cannot be a violation of his or her liberty, since the person at stake does not yet exist.[8] Therefore, he finds Habermas' claim, that an intervention into the natural genesis of a person destroys his or her liberty, unconvincing. Taking as a premise that rights serve the safety of human needs, interests and values, he disputes the "right to inherit a genetic pattern which has not been artificially changed" as postulated by the Parliamentary Assembly of the Council of Europe in the recommendation 934 (1982, 13), and claims that we cannot plausibly assume a human interest to be born with a natural disease that could have been avoided via germline intervention. He goes on to mention some important functions of the concept of *naturalness* in moral reasoning (in general, to set limits for responsibility and rationality) and concludes that, despite some connections between intuitions about *nature* and risk-assessment, *naturalness* is never a strong argument, but that risk-assessment can provide one.

---

7 In his lucid and systematic study "naturalness", Birnbacher develops and employs that twofold notion of *nature*: "'Natural' in the genetic sense is what has a natural origin, 'natural' in the qualitative sense is what does not differ from that which is to be found in nature" (Birnbacher 2014, 7). His description of these terms as 'historical terms', and his remark that one term is dependent on the other, is very helpful to understand his position, since it expresses the foundational role of a *genetic* notion of nature for Birnbacher's concept: "Taken in their genetic sense, naturalness and artificiality are historical terms; they are related to the past. […] It will not have escaped the attentive reader that the two distinct meanings of 'natural' and 'artificial' are not entirely independent of each other. […] The reason for this is that the qualitative pair of opposites is determined from the onset by their relationship to the genetic."

8 Parfit (1984), who is well known for his reductive account of personal identity.

2.1.2 Eberhard Schockenhoff

Schockenhoff first shows the narrowness of a genetic notion of *nature* by indicating that a genetic notion of nature is not, whether from a theological perspective, or from the perspective of natural philosophy, without alternatives. There are important approaches in theology that treat *creation* as an open process and man as a *created co-creator*, so that *creation* and *nature* are not something completed, or terminated at a certain point of time, but an ongoing process that includes human activity.[9] He goes on to focus on *human nature*, claiming that it is only because of the anthropological premise that human beings are endowed with reason and liberty, that we can expect justifications for actions that transform *nature*. Contra Birnbacher, Schockenhoff contends that arguments drawing on *naturalness* do have an important role to play in the debate about germline editing. He suggests a theory of personhood, that treats the human body (*Leib*) as an integrative part of the person (*leib-seelische Einheit der Person*) and as the natural precondition for his or her personal development. He claims that his defence of a "right to inherit a genetic pattern which has not been artificially changed" is not based on the moral dignity of the natural act of creation, but on an argument that spells out the idea of negative liberty: Being created in a natural way means that a person is free from the interventions of third persons and can develop in a way unconditioned by others, being conditioned only by his or her *own* natural endowment. If someday the procedure should be considered safe enough, and if unintended side-effects could be excluded, it might be justified to intervene – but only in case of severe diseases (presupposing a limited, normative concept of *disease*).

2.1.3 What Can We Learn from the Different Positions?

The foregoing presentation of two different positions on the role of *naturalness* in the debate about genome editing demonstrates that critics of the category of naturalness often work with a notion of *nature*, that can be described as genetic and totalizing.[10] Birnbacher's reference to Mill shows that the philosophical critique of this concept of *nature* and its use in normative contexts has a long tradition. Both the diagnosis of conceptual unclarity, and the idea to avoid the terms *nature*

9 For different notions of *creation* in theology, see also Dabrock (2009, 50), who explains that, in theology, "'creation' is not primarily linked to the past", but is rather a "connection between past and future, of origin and redeeming eschatological fulfillment."

10 Cf. Keil's claim about the *Verfall* (decline/decay) and *Totalisierung* (totalisation) of the concept of *nature* (Keil 1993, 360–361).

and "natural" were already formulated by Mill[11] and date back to the rise of early modern natural science to names such as Christoph Sturmius and Robert Boyle.[12] Schockenhoff's position, on the other hand, shows that this genetic notion of *nature*, as a terminated creation, is neither the dominating view in theology, nor (and here he agrees with Birnbacher and with Mill's older criticism) is it very convincing in a logical, conceptual sense. Also, by using the concept of a distinct *human nature* in an argument for moral responsibility and the need for justification, Schockenhoff employs a certain normative notion of *nature* that can be considered as an alternative to the totalizing, unstructured view of *nature*. The idea of a distinct *human nature* with the distinct ability of self-transformation implies the notion of nature as a structured whole, that contains different entities within itself, such as humans, animals, plants, and so on.

2.2 The Case against Nature: the Politics of Bioliberals in Bioethics

Birnbacher's arguments against the contested use of a genetic, totalizing concept of *nature* are very convincing. Since (among others) John Stuart Mill already formulated similar, convincing critiques of these kind of arguments in 19th century Britain, and since, as we have indicated, the genetic, totalizing notion of *nature* is far from being without alternatives, we might ask why current analytical bioethics focuses so intensely on the refutation of arguments from *naturalness*, often by drawing on conceptual problems.[13] I share an interest in this question with Düwell (2013, 126), who points out:

> "What is surprising is the fact that the refutation of arguments from naturalness takes up such a large place in the bioethical debate. [...] In many debates one gets the impression that it is very convenient to have pushed the opponent into the 'naturalness-corner'. If critics of genetic and biotechnology have nothing more to offer against the technological changes than the reference to their unnaturalness, then it is easy to dispose of one's reservations simply by pointing out that people have always

11 "The words have thus become entangled in so many foreign associations, mostly of a very powerful and tenacious character, that they have come to excite, and to be the symbols of, feelings which their original meaning will by no means justify, and which have made them one of the most copious sources of false taste, false philosophy, false morality, and even bad law" (Mill 1985, 373).

12 For references and a discussion of that trend, see Spaemann (1973, 956).

13 Cf., instead of a large list, Moriarty (2013) and Lustig (2013) and the literature they discuss.

changed nature around them and even their own nature, and that, therefore, the new technologies are not introducing any truly radical change, but merely represent further strides on well known cultural paths. In any case, it seems to me that the frequency with which the arguments from naturalness are criticised calls for illumination just as much as the lack of willingness to understand what people intend to express with these arguments in the first place."

In my view, the surprising frequency and vehemence of these efforts can be explained, at least in part, by political reasons. In these *cases against nature*, we find a puzzling mixture of conceptual and political arguments, an entanglement of the perspectives of politics and philosophy. In his study "Naturalness", Birnbacher summarizes different reasons why a concept of *human nature* could not be used in a normative context to "serve as a guideline for the future self-designs of the species" (Birnbacher 2014, 160). Next to the well known argument that a genetic, historical notion of *nature* contains conceptual problems, Birnbacher highlights political reasons – or, to be more precise, the supposed political functions of the concept – to criticize it.

First, there is the assumption and fear that claims about *human nature* could be taken as standards of normality, and thus, motivate political discrimination.[14] So, the concept is criticized with respect to its supposed political, namely undemocratic, illiberal functions.[15]

Second, Birnbacher's evaluation and critique of the concept of *human nature* addresses other *political* functions and perspectives: He criticizes the concept because, due to its normativity, it cannot "solve differences" and is thus not "useful" for pluralistic, multicultural societies that contain different conceptions of *human nature*:

"An appeal to "human nature" is not capable of resolving such differences. The image that each of the disputing parties has about humans is, from the outset, too strongly shaped by their respective normative positions. In this respect and in terms of the human species, naturalness is not a particularly useful category" (Birnbacher 2014, 162–163).

---

14 Cf. Birnbacher 2014, 161: "The attempt to make the history of mankind up to the present day the standard of human nature involves not only epistemic but also moral risks. The reason for this is that statements concerning human nature literally invite to be normatively understood and to act as standards of 'normality'– with the result that characteristics or behaviors that do not correspond to this nature are subject to discrimination."

15 There are, however, important positions that argue convincingly "that there is no conceptual or empirical necessity that species members lacking features of the characteristic life form [...] need be the victims of exclusion" (Roughly 2008, 28).

Third, there is an astonishing elitist and possibly undemocratic prejudice to be found in some criticisms of arguments from *nature*. Düwell rightly suggests that the frequency and ambition surrounding efforts to criticize arguments from *nature* might result from elitist prejudices (Düwell 2013, 126). And in fact, we do find tendencies like these in many of these criticisms. At the end of his book Birnbacher follows, or at least quotes, Dworkin's famous appeal to courageously play with fire and, in this instance, conveys Dworkin's image of "'the irresponsible cowardice' of *the many*" (Birnbacher 2014, 167, emphasis by E.O.).

Fourth, one gets the impression that Birnbacher's philosophical standpoint is akin to a political confession. Despite his previous claim that concepts of human *nature* cannot serve as a guideline for the future self-designs of the species (quoted above), and despite his argument that we should avoid appeals to human *nature*, since these are normative concepts (see above), both Birnbacher's arguments against the utility of a concept of *human nature* and his arguments *for* a right to self-modification are themselves based on a normative concept of *human nature*. From the anthropological premise that the essence of humans consists in being self-transformative beings, Birnbacher argues against the possibility of employing the concept of *human nature* as a guide for future changes in the species:

> "A second reason for the assumption that "human nature" in its broadest sense *cannot seriously be considered as a guide for future changes in the species* is the above-mentioned fact that the (successful) attempt to exceed his own nature and the natural environment, constitutes a basic and "typical" characteristic of the human species. If there is something which is "*essential*" to humanity, then it is the *ability and willingness for self-transformation* […]. The essence of man includes, among other things, the ability not only to define his individual "being," but also to create his mode of existence in accordance with that definition" (Birnbacher 2014, 161, emphasis by E.O.).

This essentialist notion of a human *nature* is used by Birnbacher in a normative way to describe the dignity and liberty of humans:

> "Even if one does not go so far as to say, in the words of David Heyd, that "the unique value of humanity – its dignity – lies in its power of self transcendence, of being other than the natural given" (Heyd 2003, 168), it can still be stated, that if humanity as a whole is to be assigned dignity, it is at least also due to its *capacity for self-design*. The freedom of man as being "freed from creation" lies precisely in *his freedom to make* his own *nature* the *object of a planned design*" (Birnbacher 2014, 162, emphasis by E.O.).

But if the concept of a *human nature* should not, according to him, be used in bioethical discussions of pluralistic societies, because of its normativity, why then does he ground his own "secular humanist" position on a 'confession' to just such a normative concept of *human nature*, using that normative concept in his argument for a "right to self-perfection"?

> "Yet a *secular humanist*, which the present *confesses* to be, will be taken aback by the almost metaphysical vigor with which many bioconservatives negatively evaluate the further implementation of technology for the good of man [...]. [...] The secular humanist, on the contrary, *welcomes every attempt* to mitigate the contingencies and dependencies of humans on external nature, but also on his own nature and *to strengthen his autonomy*; not just through education, but *also through the development of technology and medicine*. But even if we do not go as far as [Mill, E.O.] to demand a duty of self-perfection (to whom? one might ask), one cannot, however, deny human beings their right to self-perfection (Birnbacher 2014, 163, emphasis by E.O.).

If the "essence" of the human is seen "in his freedom to make his own *nature* the object of a planned design", and if it is taken as a basis for the positive evaluation of "the further implementation of technology for the good of man", is Birnbacher not then doing exactly what he himself previously criticized, namely using a normative concept of *human nature* in bioethical discussion? Is the premise that nature can and should be dominated, controlled, and made "the object of planned design" not normative and political in the sense that it presupposes the right to dominate, control and use nature? In his chapter on "The Naturalization of Human Dignity", Birnbacher first reminds us of the *philosophes*, and Pico della Mirandola, to argue for a principle of human dignity that is based on autonomy and self-perfection.[16] He uses this reconstruction to argue against some positions, that, following Kant and Hegel, bestow dignity on the human body and treat it as a "subject of rights." He goes a step further with Kant, conceding that the following premise

> "is plausible in itself, that the human being, who is ascribed the postulate of human dignity, [...] is not a pure spirit but instead an embodied, psychophysical being. The dignity that is proper to him, regardless of what this consists in, not only adheres to his intellectual, but also to his physical, material existence" (Birnbacher 2014, 117).

Although Birnbacher afterwards severely criticizes Kant's conclusions, claiming that a "naturalization" or "biologizing" of human dignity in the Kantian tradition is "highly problematic" and "subject to serious objections", Kant is, nevertheless, used a few lines later as an authority, who testifies that "the special status of man,

16 "The philosophes, the pioneers of the French Revolution, even Pico della Mirandola in the Renaissance, viewed human dignity primarily with regard to the human being's autonomy and his abilities, if not even in his obligation to self-perfection" (Birnbacher 2014, 117).

which is accentuated by the postulate of human dignity, is located in his spiritual *nature* and not in his biological-material nature" (Birnbacher 2014, 119). Taking up his former activity of drawing a historical line from Pico della Mirandolla and the *philosophes* to Kant, Birnbacher builds a *genetic* argument that works with a certain *historical* notion of human dignity:

> "The postulate of human dignity implies neither duties to oneself nor duties to our own body. Whether or not duties to oneself can ever establish primary and non-derived duties is one thing, but they certainly cannot be derived from a principle that *has developed historically* as a principle of freedom and not as one of duty" (Birnbacher 2014, 120, emphasis by E.O.).

Is an argument that tries to justify a certain notion of human dignity via the *historical development* of that principle[17] not on the same level as the much-criticized arguments that work with a genetic (historical) notion of *nature*? (In both cases, the authority of the old, it seems, is meant to replace a thorough philosophical argument which justifies *why* something is good.) To me at least, the contours of Birnbacher's theory of personhood, and the role the body plays in it remain highly unclear after this chapter – given his acceptance of the premise, that man is an "embodied psychophysical being", and that dignity adheres "also to his physical, material existence", and his insistence that "the metaphysical foundations of this principle […] are freedom, reason and moral capacity" (Birnbacher 2014, 119). Or is the lack of clarity about the body's role in this concept of personhood a result of the (philosophically nettlesome) proposal to avoid the concept of the person in bioethics?[18]

Whatever the case, at least we have found some plausibility for our claim that refutations of the concepts of *nature* and *naturalness* are sometimes (also) motivated by political considerations. These political considerations unveil a conception of democracy that is elitist on the one hand, and (in strange ways) separated from normativity (and normative reasoning) on the other hand. We (who potentially belong to the crowd of irresponsible cowards) are told that the authorities (Pico della Mirandola and the *philosophes* of the French revolution) were right, and that Kant was right (though, not in his treatment of the human body), and that Kantian philosophy and German constitutional law testify to the metaphysical foundations of the principle of human dignity. We are presented with the conclusion (a political

17 I am not sure that all human rights historians would agree with that claim about the historical roots of that principle.

18 This proposal is articulated in Birnbacher's book, "Bioethik zwischen Natur und Interesse" (2006, 73–74).

goal), that according to this principle, liberal societies "must allow changes by which these constitutive capacities can be enhanced" (Birnbacher 2014, 119). However, we are left wondering *why* Pico, Kant and the *philosophes* were right and why we should accept their position.[19]

2.3 The Case for Nature: The Political Dimension of (Human) Nature and the Necessity of a Political Theory of the Human Body

In the end, then, I do not see that bioliberals succeed in their proclaimed goal of avoiding concepts of human nature. *Bioliberals* (if we provisionally accept that over-simplification) are right to point out that arguments declaring historically developed phenomena to be *natural*, with the intention of shielding them from critique in absence of thorough arguments, are not very convincing. They are right to criticize the strategy of "appealing" to *nature* as one of immunization against philosophical critique. On the other hand, however, *bioliberals* and transhumanists themselves do use normative concepts of *human nature*, as a foundation for their normative conclusions, (sometimes without thoroughly arguing) for it. While *bioliberals* typically put a strong emphasis on the principle of neutrality,[20] they themselves use normative conceptions of *human nature* as a basis for their normative conclusions. Since a common critique of bioliberals is that bioconservative positions are unscientific or dogmatic, it might surprise that the bioliberals' normative conceptions of human nature are often stated as a *confession* (Birnbacher) or *belief* (Savulescu):

> "I *believe* that to be human is to be better. Or, at least, to strive to be better. We should be here for a good time, not just a long time. Enhancement, far from being merely permissible, is something we *should* aspire to achieve" (Savulescu 2007, 533, emphasis by E.O.).

Is not this *belief in* and *confession of* a certain normative conception of *human nature*, and its (unjustified) use for normative conclusions about biopolicy in liberal

19 In this chapter, Birnbacher develops arguments *against* the conservatives who follow Kant in treating the human body as a separate entity. But I can see no arguments, except the historical ones, that argue positively for a certain interpretation of human dignity, and I would be interested in an explanation of the role of the human body in his theory of personhood.

20 Savulescu 2007, 526: "It is a basic principle of liberal states like the United Kingdom that the state be 'neutral' to different conceptions of the good life."

democracies, problematic from a philosophical perspective and at odds with the so called "principle of neutrality"?[21] Are these not normative and political *appeals*, rather than arguments?

Nida-Rümelin recently formulated a thorough critique of a certain way of using anthropological arguments that is of utmost importance for our question about the role of conceptions of human *nature* in philosophical and public debates. According to him, ethical fundamentalism ("ethischer Fundamentalismus") and epistemological foundationalism ("erkenntnistheoretischer Fundamentalismus") both have a tendency to shield anthropological premises from critique, either by dogmatically declaring them the untouchable essence of a certain society's identity, or by declaring them to be the foundation of argumentation which is not capable of, or dependent on, further justification (cf. Nida-Rümelin 2015, 6–7). He argues convincingly that an argumentative strategy that tries to avoid anthropological premises (or, following our argument, tries to avoid normative conceptions about *human nature*) is impossible: We cannot decide on being normative or not, (or on using normative conceptions about *human nature)*, but, as living and acting human beings (for example, writing as philosophers about *genome editing*), we always *follow and express* normative conceptions about human *nature*.[22]

I want to follow this argument and link it with the hypothesis that our relation to *nature* in general, and to our own body in particular, is normative and essentially *political*: We cannot opt to have no relation at all. Rather, as *embodied*, living, moving beings endowed with reason, we always take a position *with our bodies*, towards other bodies and towards nature. More important, as living bodies endowed with reason we have, from the very beginning, a normative relation *to our own living body* as a distinct part of *nature*.[23] The history of political ideas can sensitize for the fact that modern natural science is not metaphysically or politically neutral in any meaningful sense, but rests on a fundamental change in our normative relation to

21 Even John Rawls concedes that liberalism cannot be "neutral" in a comprehensive sense, concluding that this is a reason to avoid that very term: "Neutrality of effect or influence political liberalism abandons as impracticable, and since this idea is strongly presupposed by the term itself, this is a reason for avoiding it" (Rawls 2005 and 1993, 194). Cf. Kauffmann 2001, 195–215, esp. 202–204.

22 "Anthropologie ist also nichts, wozu ich mich entscheiden kann: Nehme ich Stellung oder nicht? Sondern wir nehmen immer schon implizit Stellung." (Nida-Rümelin 2015, 11). "Anthropology is not something on which I may decide whether or not to take a position. Rather, we always already implicitly take a position."

23 Therefore, it is my contention that aspects of both Nida-Rümelin's concept of *Lebenswelt* and of Plessner's concept of *exzentrische Positionalität* would be integral parts of a theory of the human body that aims to spell out its political dimension.

nature. Bacon, Hobbes and Descartes paved the way for a conceptualization of nature as a dead, meaningless matter that can be dominated, controlled and possessed. Hobbes established a sort of "inverted teleology" that acknowledged the mortality of the living human body as political important, but fits in the overall tendency to get rid of teleology.[24] The new conceptualization of the world in a materialistic and mechanistic way served the new notion of science as a means to increase human power: For Bacon, science served the practical end "to command nature in action,"[25] for Descartes, man should become "maître et possesseur de la nature" (Discours de la methode, VI 3). The renunciation of a teleological perspective in modernity and the new notion of a materialistic, mechanistic world fulfil political functions, namely the justification of relations of domination, appropriation and control with respect to external nature. For contact with the natural human body this has grave consequences: The disrespect of the independence of the living, self-moving body of another as a limit necessarily represents a violent action, as Hegel still recalled and warned: "To apply violence to my body is to apply violence to me" (Elements of the Philosophy of Right, §48). This consciousness for the independence of nature (and in particular for living human embodiment) is successively pushed back via the paradigm shift of early modern natural science and is also not always present in contemporary liberal societies. It represents an astounding irony that liberals, who had ever strictly respected the protection of the individual person from interventions via governments or fellow humans, have frankly recommended invasive assaults and interventions in and upon the body and the genetic composition of other human beings.[26] While I do not agree with claims that bestow prudential value (cf. Schramme 2011, 79) or normativity (Siep 2005; see also Reichhold and Delhorn 2011) to the body itself, I think the possibility and necessity of humans putting themselves in a justified, normative *relation* to their bodies is of utmost importance for understanding and practicing normative reasoning and political action:[27] With our bodies, we are confronted with a part of nature that both enables and limits our personal space of liberty as rational beings and thus with the task to shape a relationship that is, therefore, deeply political. With our bodies, we can experience the conditions of biological life (e. g. biological needs, hormone level, conditions of healthiness), and we are constantly confronted with the question of how to relate to that *independency* of life – both with respect to our own bodies,

24 See Odzuck 2015.

25 Bacon 2000, 15–16.

26 Karnein 2013.

27 I'll explore these hypotheses in my second book project in which I plan to develop a political theory of the body.

and with respect to the living bodies of other humans, (that we can learn to see *as independent living beings*, and not (only) as physical hindrances, by a reflection on our own nature). Core political concepts for liberal democracies, such as dignity, violence and liberty, are tied to the notion of a living human body and presuppose to explicate and to enfold analytically normative claims about *human nature* and the normative relation to the human body.

Avoiding normativity, or normative claims about *human nature*, is thus neither a real possibility for living human beings, nor for their life in a liberal democracy. While we cannot opt to avoid normativity, we can choose the degree of reflection and justification that surrounds and grounds our normative conceptions. Since liberal democracies rest on distinct, normative anthropological premises, the *life* and quality of liberal democracies rests and depends on the articulation *and* justification of its own anthropological premises. People can only be expected to act in a way that accords with the fundamental norms of liberal democracy if they know, understand, share and are able and willing to justify and defend those norms.[28] While keeping *nature* and concepts of *human nature* out of liberal democracies is impossible,[29] bringing nature back in[30] could work, especially if we work with the premise that (whatever *nature* in general, and our human nature in particular, is) *nature* is both malleable and consists of different "selves",[31] and therefore, every decision to act or not act, and thus, to treat or not to treat *these independent entities* in a certain way is part of a normative and political relation to *nature* that has to be justified with good reasons.[32]

As part of a *malleable nature*, humans (as malleable creatures) possess the ability to think about the desirability of changing (them) selves. Nida-Rümelin argues convincingly that reason, liberty and responsibility belong together and can be considered important elements of a normative anthropology (Nida-Rüme-

28 On the normative and anthropological foundations of democracy, compare Nida-Rümelin (2015, 9.14–16) and Kauffmann (2011, 10).

29 Compare also Brooke's critical answer to the question whether the word *nature* should be eliminated: "'[N]o amount of well-intentioned intellectual political hygiene' is likely to get rid of it." (Brooke 2009, 329).

30 In contrast to recent approaches that bring *nature* back in like evolutionary ethics, environmentalism and neo-Aristotelianism, my approach does not claim to be naturalistic, but rather, a normative political philosophy. For an overview over recent "approaches that bring nature back in" (Zammito et al. 2008, 122–125).

31 Lustig 2009, 240: "[T]he vision of malleable nature [...] foregrounds questions about the normative warrants for human alterations of nature."

32 A similar argument is indicated by Schockenhoff, see above. Thanks to Clemens Kauffmann for helpful conversations about teleology and being a self.

lin 2015, 14 and his other publications mentioned there). In my view, we should also spell out what it means to be an *embodied*, living being, and to consider the living body (and its dependency and independency) as necessary ingredients of such a normative, political anthropology. Such a normative anthropology would have to develop arguments for why we should accept a self-description different to Hobbes's narrative of isolated passion-driven bodies moved by natural laws.[33] A political theory of the body that can illuminate the special qualities of humans, as moved *and* living, self-moving bodies, and spells out the political dimensions of our relations to our bodies is, in my view, necessary for a responsible biopolicy in liberal democracies.

3 Conclusion

Coming back to our original question, we can now say that the question whether genome editing *is unnatural* might be the wrong question. *Bioliberals* are right that simple *appeals* to *nature* are not convincing. However, they are wrong when (in the name of a "principle of neutrality" or out of fear that the concept of *nature* is *per se* tied to an illiberal attitude) they avoid the task of developing and justifying a normative anthropology. We should indeed ask ourselves if we have good reasons to intervene into the human genome and into the constitution of future persons. These reasons, however, are necessarily connected with a normative concept of nature and of *human nature*, and it is irresponsible to blur that connection or to avoid that question. A responsible policy on the human genome, and a responsible biopolicy in general, will depend on our efforts to develop and justify normative concepts of nature[34] and of *human nature*, and especially to spell out and justify our normative and political relation to the living human body.

33 On Hobbes's body-based liberalism, compare my first book (Odzuck 2015).

34 Although the premise that concepts of "nature" and "naturalness" can fulfill important functions in philosophical arguments is opposed to Birnbacher's position, any conceptual work in this area will profit enourmously from his lucid and systematic treatment of these concepts in his book "Naturalness".

References

Bacon, Francis. 2000. *New Organon*. Cambridge: Cambridge University Press.

Baltimore, David, et al. 2015. A Prudent Path Forward for Genomic Engineering and Germline Gene Modification. *Science* 348: 36–38.

Birnbacher, Dieter. 2006a. *Bioethik zwischen Natur und Interesse*. Frankfurt: Suhrkamp.

Birnbacher, Dieter. 2006b. *Natürlichkeit*. Berlin: De Gruyter.

Birnbacher, Dieter. 2014. *Naturalness. Is the "Natural" preferable to the "Artificial"?* Lanham: University Press of America.

Bostrom, Nick. 2005. In Defense of Posthuman Dignity. *Bioethics* 19: 202–214.

Brooke, John H. 2009. Should the Word "Nature" Be Eliminated? In *Envisioning Nature, Science and Religion*, James D. Proctor (ed.), 312–336. West Conshohocken: Templeton Press.

Clark, Stephen, and Rebecca Roache. 2008. Enhancement am Menschen, Intuitionen und die Weisheit des Nachdenkens über den Widerwillen. In *Der neue Mensch? Enhancement und Genetik*, Knoepffler, Nikolaus, and Julian Savulescu (eds.), 55–81. Freiburg/München: Karl Alber.

Dabrock, Peter. 2009. Playing God? Synthetic biology as a theological and ethical challenge. *Systems and Synthetic Biology* 3: 47–54.

Dabrock, Peter. 2017. Fundamentale Eingriffe in das menschliche Erbgut. Und niemand spricht darüber. *Süddeutsche Zeitung*, August 30.

German Ethics Council. 2016. *Zugriff auf das menschliche Erbgut. Neue Möglichkeiten und ihre ethische Beurteilung. Jahrestagung am 22.06.2016*. Simultanmitschrift. http://www.ethikrat.org/veranstaltungen/jahrestagungen/zugriff-auf-das-menschliche-erbgut. Accessed 15 October 2017.

German Ethics Council. 2017. *Ad hoc Recommendation: Germline Intervention in the Human Embryo: German Ethics Council Calls for Global Political Debate and International Regulation*. http://www.ethikrat.org/publications/ad-hoc-recommendations/keimbahn%20eingriffe-am-menschlichen-embryo. Accessed 15 October 2017.

Düwell, Marcus. 2013. *Bioethics. Methods, Theories, Domains*. London/New York: Routledge.

Dworkin, Ronald. 1999. *Playing God*. http://www.prospect-magazine.co.uk/magazine_/playing-god-genetics-ronald-dworkin. Accessed 15 October 2017.

Karnein, Anja. 2013. *Zukünftige Personen. Eine Theorie des ungeborenen Lebens von der künstlichen Befruchtung bis zur genetischen Manipulation*. Berlin: Suhrkamp.

Kauffmann, Clemens. 2001. „Clash of Views": Was fehlt dem politischen Liberalismus zur Moral des 21. Jahrhunderts? In *Werte im 21. Jahrhundert*, Walter Schweidler (ed.), 195–215. Baden-Baden: Nomos.

Kauffmann, Clemens. 2011. Biopolitik im liberalen Staat. Einleitung. In *Biopolitik im liberalen Staat*, Clemens Kauffmann (ed.), 9–12. Baden-Baden: Nomos.

Keil, Geert. 1993. *Kritik des Naturalismus*. Berlin/New York: De Gruyter.

Lanphier, Edward, et al. 2015. Don't Edit the Human Germline! *Nature* 519/7544: 410–411.

Lustig, Andrew. 2009. The Vision of Malleable Nature: A Complex Conversation. In *Envisioning Nature, Science and Religion*, James D. Proctor (ed.), 227–244. West Conshohocken: Templeton Press.

Lustig, Andrew. 2013. Appeals to Nature and Naturalness in Debates about Synthetic Biology. In *Synthetic Biology and Morality. Artificial Life and the Bounds of Nature*, Cabnick, Gregory E., and Thomas H. Murray (eds.), 15–34. Cambridge: MIT Press.

Mill, John Stuart. 1985. *The Collected Works of John Stuart Mill. Volume X – Essays on Ethics, Religion, and Society*, John M. Robson (ed.), Toronto: University of Toronto Press, London: Routledge and Kegan Paul.
Moriarty, Paul V. 2013. Nature and the Natural. In *The International Encyclopedia of Ethics*, Hugh LaFollette (ed.), 3549–3558. Hoboken, New Jersey: Wiley-Blackwell.
Nida-Rümelin, Julian. 2015. Plädoyer für eine normative Anthropologie. In *Anthropologie und Ethik*, Heilinger, Jan-Christoph, and Julian Nida-Rümelin (eds.), 3–18. Berlin/Boston: De Gruyter.
Nuffield Council on Bioethics. 2015. *Ideas about Naturalness in Public and Political Debates about Science, Technology and Medicine. Analysis paper.* http://nuffieldbioethics.org/project/naturalness/the-findings. Accessed 14 September 2017.
Odzuck, Eva. 2015. *Thomas Hobbes' körperbasierter Liberalismus. Eine kritische Analyse des Leviathan*. Berlin: Duncker und Humblot.
Parfit, Derek. 1984. *Reasons and Persons*. Oxford: Oxford University Press.
Rawls, John. 1993. *Political Liberalism*. New York: Columbia University Press.
Reichhold, Anne, and Pascal Delhorn (eds.). 2011. *Normativität des Körpers*. Freiburg: Karl Alber.
Roughley, Neil. 2011. Human Natures. In *Human Nature and Self Design*, Sebastian Schleidgen, et al. (eds.), 11–13. Paderborn: Mentis.
Savulescu, Julian. 2007. Genetic Interventions and the Ethics of Enhancement of Human Beings. In *Oxford Handbook of Bioethics*, Bonnie Steinbock (ed.), 516–535. Oxford: Oxford University Press.
Schiemann, Gregor. 2011. Naturalness and Artificiality in Bioethics. In *Human Nature and Self Design*, Sebastian Schleidgen, et al. (eds.), 99–109. Paderborn: mentis.
Schockenhoff, Eberhard. 2013. *Ethik des Lebens. Grundlagen und neue Herausforderungen*. Freiburg: Herder.
Schramme, Thomas. 2011. The Body as Source of Prudential Value. In *Human Nature and Self Design*, Sebastian Schleidgen, et al. (eds.), 67–81. Paderborn: mentis.
Siep, Ludwig. 2008. Normative Aspekte des menschlichen Körpers. In *Die menschliche Natur. Welchen und wieviel Wert hat sie?* Kurt Bayertz (ed.), 157–173. Paderborn: Mentis.
Spaemann, Robert. 1973. Natur. In: *Handbuch philosophischer Grundbegriffe. Band II. Gesetz – Relation*. Hermann Krings, et al. (eds.), 956–969. München: Kösel-Verlag.
Zammito, John H., et al. 2008. Philosophical Approaches to Nature. In *Altering Nature. Volume I: Concepts of 'Nature' and 'The Natural' in Biotechnology Debates*, Lustig, B. Andrew, et al. (eds.), 63–136. Heidelberg: Springer.

Transition and Care

Theological Concepts of Dynamic Creation and the Ethics of Genome Editing

Mathias Wirth

Keywords

Agency, Nature, Process Theology, Theological Ethics, Theology of Creation

Abstract

Genome editing, or gene surgery, demands a theological and ethical exploration, as the applied methods that rapidly change the genetic make-up of humans seem to be making deep interventions. This incites theological and ethical discussion since modifications of the germline, a very fundamental dimension of the bodily sphere, seem to somehow contradict what is considered, from a theological perspective, a good creation. The active transformation of gene sections, by inserting or extracting information, constitutes a new dimension of gene manipulation, as the term "genome editing" indicates. There has already been theological and ethical scrutiny of the arguments, addressing positive claims for such interventions and providing an alternative to the playing-God-argument that is always brought up in the face of new developments, especially in technology-based medical bioscience, and is concerned with all sorts of deep changes of the body. This contribution aims to add affirmative arguments, from a Christian theological understanding in dialogue with process theology, to the debate on genome editing and to the transition of the biological basis of human beings.

M. Braun et al. (Eds.), *Between Moral Hazard and Legal Uncertainty*, Technikzukünfte, Wissenschaft und Gesellschaft / Futures of Technology, Science and Society, https://doi.org/10.1007/978-3-658-22660-2_9

„Wenig Wissenschaft entfernt von Gott,
viel führt zu ihm zurück."[1]
Louis Pasteur (1822–1885)

Introduction

In 2015, the German Research Foundation (Deutsche Forschungsgemeinschaft) declared that not all forms of genome editing were ethically questionable. On the contrary, gene insertions, gene deletions and complete gene knockouts in cell types, in order to alter a certain *status quo*, had already become common practice. For instance, applying CRISPR-Cas9 as a tool in experimental models was considered a "quantum leap" in genetic engineering (Baltimore et al. 2015, 36). Cutting gene sequences with "user-friendly technologies" (the affordable "molecular scissors" for changing DNA) could be performed for different reasons in the future: to disable or repair genes, and to replace parts of genes with a new strand of DNA. All this can be done with therapeutic intentions, but it also catalyzes the options and aspirations of human transformation in terms of human and environmental betterment and enhancement (Baltimore et al. 2015). The therapy-argument is a strong one (Kipke et al. 2017) in the ethical debate, whereas affirmative or negative arguments on the general human potential for transformation seem weaker in a cultural setting where enhancement does not seem to be regarded with as much importance as curing diseases. Several questions regarding genomic engineering remain unresolved and many concerns persist: Unintended and hereditary modifications, unpredictable reactions of the gene environment, unclear pathways of diseases and individual expression are some core issues (Baltimore et al. 2015). Considering the great number of risks, these concerns sum up the potential dangers of the new gene manipulation technology. However, as some argue that since genome editing is seen as one of the most promising technologies "to reshape the biosphere for the benefit of the environment and human society" (Baltimore et al. 2015, 36), possible side effects can be seen as acceptable "when the reward of success is high" (Baltimore et al. 2015, 38). This robust utilitarian standpoint alludes to a general pattern of therapeutic strategy that accepts possible side-effects and harm when the reward is high.

This contribution focusses on the anticipated option of transforming individuals' genetic configuration – for purposes that are not purely therapeutic – through enhancing gene editing, casting light on some arguments from theological ethics

1 "A little science estranges one from God, but much science leads one back".

that are crucial for adopting a coherent and informed position on genome editing. One starting point is found in the observation that transformation belongs to the history of nature. Neither earth nor humanity nor God must be seen as unchanging, static collections (Whitehead 1977). Moments of transformation can be observed in many contexts and, though some transformations, like trans-sexuality and trans-humanity, may provoke reservation or angst, since both concepts seem to threaten fundamental beliefs about gender, biology, and humanity in general, the ideas of trans-cendence, trans-nationality, trans-politics or any kind of trans-ferring, tend to have positive connotations (Wirth 2018a). Many transitions seek to meet the human need for an overall betterment of human nature *(Mängelwesen)*, a need to which gene editing also alludes. The religious concepts of transcendence and the biblical 'new human' are deeply internalized analogies to the general wish for profound transition and technical modification for self-betterment. Thus, there should not be a general opposition to ideas of transition amongst Christian theologians. An example of a transition-positive argument is the newer interpretation of the *dominium terrae* (Gen 1:28), which is widely accepted as allowing therapeutic interventions in a world that is different from God and may be manipulated (Dabrock 2009). In more general, theological terms: The *status quo* of nature, including human nature, is not the aim of creation. Taking care of creation is therefore not always equal to conserving it (Koslowski 2002).

It is with this in mind that this contribution focusses on the wide vista of options for transition that can be anticipated for genome editing. Opening a conversation with proponents of a dynamic take on the theology of creation (for example process theologians), who join an intensive dialogue with science, is intended to cast some light on the possible positive theological value of the transformations (for therapeutic and non-therapeutic purposes) that may accompany the development of genome editing technologies. An analysis and application of the works of Alfred North Whitehead, Ian Graeme Barbour, and Arthur Peacocke, who all contribute ideas of a developing creation, shape the methodology of this contribution.

2 Alfred North Whitehead and the Quest for Perfection

The British philosopher and mathematician, Alfred North Whitehead (1861–1947), is famous in philosophy and theology for his concept of process philosophy which is inherently rooted in and thus stimulates theology. His philosophical findings are widely recognized (Rohmer 2000) and are mostly applicable to philosophical as well as theological questions. However, it should be noted that his *œuvre* is highly complex and difficult to access. One reason for this is the complexity of the questions, for example, where he intends to explicate an entire cosmological concept. Another reason is his artificial language, so full of neologisms that a Whitehead-glossary becomes necessary (Müller 2009).

The unifying theme of Whitehead's work is his study of the process-character of nature. Being in processes, according to Whitehead, means being in a state of flux and is thought of as a basic characteristic of reality (Müller 2009; Sander 1991; Schramm 1991). This development appears as a continuous transformation, and Whitehead understands this to be the innermost nature of reality, governed by the principle of novelty (Rohmer 2000). This presupposes the ability that experiences can be made. These experiences can be intensified by the implementation of high-contrast components (Griffin 2004; Hampe 1998; Müller 2009; Schramm 1991). In other words, a process can be observed when existing entities manage to let a new entity appear (Müller 2009). With this process thinking comes the conviction of a free will without any deterministic framing (Müller 2009; Schramm 1991) that does not deride human action as being part of a "divine puppet theatre" (Polkinghorne 2004, 67; Polkinghorne 1998). Consequently, God cannot be considered omnipotent in any sense that would annihilate human freedom (Griffin 2004). In this framework, freedom of action, which Whitehead defines as a "primary human need", can be implemented as relative autonomy without reservation (Whitehead 1969, 172; cf. Maaßen 1990; Rohmer 2000) and is also described in Whitehead's work as "self-realization", "self-creation", or "self-causation". In this way, Whitehead seeks to provide opposition to theories of substance *(Substanzdenken)* (Rohmer 2000). Freedom of action, however, is only possible within a non-despotic religious atmosphere, a point which exposes, *ex negativo*, the long history of restricted action based on religious beliefs in an authoritarian God (Whitehead 1969). Thus, process thinking in theology allows for a renewed discovery of human potential, and with this the potential of freedom. For this reason, Griffin judges Whitehead's theory as promoting "world-loyalty" (Griffin 2004, 290). Moreover, Whitehead asserts that his theory's impact upon ethical thinking is of no little importance. He warns against underwriting ethical codices with extreme demands and high dogmatic components, because these would tie in with the multitude of authoritarian morals from above (Whitehead 1969). Whitehead emphasizes that the criterion for an adequate ethical concept is its impact on supporting development towards perfection (Whitehead 1969). Future genome editing, both for the sake of curing diseases and for the sake of implementing desirable genes or terminating undesirable ones, could be seen as concretized development towards material and (closely linked therewith) personal betterment. The task of ethics could be to participate in scrutinizing what could foster betterment and what is likely to do the opposite.

Closely related to Whitehead's process thinking is what he terms "creativity", which he defines as the "principle of novelty". The main characteristic of creativity is seen in the production in nature of something new (Sander 1991), that has never existed before (Whitehead 1977; Rohmer 2000; Schramm 1991; Welker 1988). In this possibility of the new lies untapped potential for overcoming evil (Sander 1991), inasmuch as the *status quo* appears, at least in parts, as a deficiency, and overcoming it can allow for aspects of betterment. Admittedly, even new developments can lead to new modes of both moral and natural evil. Whitehead does not understand the material world as "passive receptivity"; therefore God will experience a world to come that was not simply there in the beginning (Whitehead 1977). This implies changes within God, a thought that provokes much reservation with regard to process thinking amongst theologians (Müller 2009; Welker 1988), even though a dynamic understanding of creation finds important links to theology in general.

Also relevant for applying some of Whitehead's thoughts to the ethical debate on genome editing are his ideas concerning the relationship between science and religion. According to Whitehead, religion and science are the two main forces affecting humanity, but both appear as contrary approaches. What they have in common, though, is the historicity of knowledge: Religion and science are characterized by a history of both knowledge and misjudgment. According to Whitehead, this has been a hard lesson to learn, especially for Christianity. Straight away, right at the point of departure, lies the first gross misunderstanding: the imminent expectation of Christ's return. Because this scenario was never realized, the early Christians had to recognize that they misunderstood Christ and had to change their eschatological expectations. That might have taught them a lesson about the necessity of a transformative faith. Nevertheless, until now, a transformative faith has been conceived as something negative, which endangers faith (Whitehead 1990).

Whitehead considers the stress placed on the conflict potential between religion and science to be overblown, since their material objects differ completely. Science is concerned with the natural law, whereas religion is concerned with aesthetical and moral values. What follows is that both miss and find certain aspects of reality; what one focusses on is something the other oversees, and vice versa (Whitehead 1990; Whitehead 1969). Admittedly, religions differ remarkably from science, as Whitehead claims, when it comes to confrontations with new developments. Common, religiously motivated reactions to new scientific findings are rejection, despair, or even anathema (Whitehead 1990). According to Whitehead, this especially results from the widespread monotheistic belief that the world is the result of God's holy decree, which must be obediently followed. But, owing to the process character of reality, this way of thinking has led and will continue to lead religion into tragedy (Whitehead 1977; Whitehead 1969).[2]

Following the aforementioned period of religious reservation, or even disgust, concerning technological improvements, it turns out to be possible to develop ample room for integrating new aspects by reinterpreting religious texts and traditions. Whitehead suggests that religions, for the sake of their survival, adopt a positive approach to new developments just as the field of science has done (Fetz 1990). Despite the timelessness of religious principles, he insists that how they are practiced must differ depending on the continuity of developments (evolutionary theory of religion) (Fetz 1990).[3] Whitehead's theory, based on the conviction of the relativity of the world and religions, appears as a major contribution to a transgressive concept of religion (Welker 1988). The modification of theology in dialogue with science should be a chief issue when religion and theology seek to participate in current controversies (Whitehead 1990). More pertinently, in times of scientific development, religions should avoid fearmongering and promoting ideas of a tyrannical God behind nature, who is easily provoked and angered by humanity for interfering with his power. Whitehead stresses that any contemporary theological concept that characterizes God in terms of power and force will very likely be rejected (Rohmer 2000; Whitehead 1990); especially by a theology sensible to theodicy. This, however, does not prevent him from painting a picture of a successfully stimulating, interfering, but also developing God in process (Hampe 1998; Rohmer 2000; Welker 1988). Hence, Whitehead's God can be understood as the "ground of novelty" and "leader of the creative advance" as well as having the

---

2 Whitehead distinguishes what he considers to be three false concepts of God: (1) God as a sovereign in the tradition of Egyptian, Persian, or Roman rulers; (2) God as a "personalized moral energy", as, according to Whitehead, present in Hebrew prophets; and (3) God as an "unmoved mover", a philosophical principle prevalent, for instance, in Aristotle. Whitehead discovers a fourth concept, enfolded in the Jesus-movement, which he welcomes, as it can be seen in congruence with his process philosophy: The "Galilean model" opposed to models of God as a "ruling emperor", "merciless moralist", or "unmoved mover", and characterized by love. Love, as Whitehead explains in line with his thinking, neither rules nor is it motionless (Whitehead 1977, 611; cf. also Fetz 1990; Griffin 2004; Maaßen 1990; Case-Winters 2004).

3 It would therefore be overly simplistic to understand the spirit of science as a "spirit of transformation" and the spirit of religion and theology as a "spirit of preservation". Whitehead presents the argument that everything would be nothing if there were no preservation at all. But this applies to science as well as religion (Whitehead 1990, 233). Moreover, it is clear that transformation and preservation are not as contrasting as they may first appear. One can take place in the name of the other, for instance religion's preservation through transformation.

role of the transformer (Case-Winters 2004, 151–152). Julia A. Lamm agrees with Whitehead that with process thinking comes an open system, and she stresses the importance of open systems in the prevention of ideologies (Lamm 2004).

Whitehead understands the current concept of evil as primarily characterized by domination and commandment, containment and violation (Whitehead 1990; cf. Maaßen 1990). The application of these findings to the bio-ethical and theological debate, and in particular to genome editing, must be read as a plea against neophobia and a warning that theological ethics could be deemed inappropriate if it perpetuates an ethic of domination, containment and violation – for instance, in the case of genome editing against persons' and future persons' (potential) suffering from genetic diseases or against people with enhancement wishes.[4] Whitehead does not suggest an anything-goes-paradigm, but favors carefully allowing for developments and transition in order to prevent catastrophes (Whitehead 1990) and to overcome evil by means of the new (Sander 1991); which ostensibly could be criticized as a certain form of over-selling the immanent future *(Zukunftsgläubigkeit)*. From this perspective, then, there is ample room to allow for biological transitions, as long as unbearable side-effects or catastrophes are excluded as much as possible (Whitehead 1990; Whitehead 1969). For an ethical debate on genome editing from Whitehead's perspective, one might ask whether new attempts to improve the human condition could be judged as concretizing inter-human care and could be correlated with an eschatological 'ethos of new creation' (Sander 1991, 282; Polkinghorne 2000). The quotation from the famous bacteriologist, Louis Pasteur (1822–1885), according to whom less science distances one from God, whilst more science brings one closer to God (Predel 1996), could be re-read ethically as a summons to discover the full potential of creation for the sake of the other.

3 Ian Graeme Barbour and the Need for the Neighbor as a Technological Imperative

As early as 1970, Ian Graeme Barbour (1923–2013), a US-American philosopher, physicist, and theologian, developed early ideas on genome editing and gene surgery. In his study on the ethics of technology in secular times, he sympathizes with

4 Whitehead explicitly criticizes what he describes as Protestant disgust of corporal aestheticism, because it seemed to measure something, the material, as valuable, although it was worthless (Whitehead 1990). Also in a wider focus, Whitehead expresses harsh criticism of Protestantism (Whitehead 1969).

process-thought in theology and stresses its capacity to affirm the factual world (Barbour 1970). Process thinking maintains that God is not the "authoritarian judge, the repressor of human vitalities" (Barbour 1970, 53). Rather, as Barbour puts it in conversation with Philip Hefner (Barbour 1990), God allows humanity to play an active part in creation in the impressive role of "co-creator" (Barbour 1970, 53).[5] This, however, leads to responsibility of action, when, according to process theology, God restricts himself strictly to a cooperative agenda and does not perform any action in isolation (Barbour 1970). This, Barbour concludes, is because God does not enforce a plan independent from human history and freedom. Accordingly, science must be seen as an agent of active reference to the world. Hence, Barbour stresses that science is not primarily about knowledge but about power (Barbour 1970). In this sense, science and Christianity find common ground, as both seek to take an active role in transforming the *status quo*. Barbour points out that this possible alliance appears to be counterfactual, as religion is regularly experienced, particularly in secular times, as suppression. Science seems exclusively to trust technological power to guarantee freedom and fulfillment, whereas religions seem to exclusively trust God's intervention and grace to guarantee freedom and fulfillment (Barbour 1970). This conflicting situation moves Barbour to focus on humankind's part in transforming the world in terms of freedom and fulfillment. In other words, does science concretize or negate how Christian theology may understand the role of humans in history and towards nature?

---

5 On the one hand, the concept of man as co-creator can be criticized due to the fundamental difference between God and humanity: "Man can principally not act like God" (Dabrock 2009, 51). In this sense, creation in a narrow or biblical sense does not apply to human action, "since the power to create (in the strict sense of the [Hebrew] *bara*) is exclusively divine and cannot be claimed by human beings" (Dabrock 2009, 52; cf. also Dabrock at al. 2010). Furthermore, John Polkinghorne articulated criticism of Philip Hefner's "created co-creator" in the same light and finds that this concept undermines the difference between humans and God and thus suggests "stewardship" as a more suitable metaphor: "To speak of ourselves as stewards acknowledges that duty of care which we owe to creation and it recognizes the Creator as the source of power and opportunity" (Polkinghorne 1996, 114-115). However, what Polkinghorne defines as care refers to a creation that, in his thinking, is understood as a processing reality, so that care does not equal maintaining any *status quo*, as could be easily misunderstood (Dinter 1999). By contrast, and from an observational perspective, there is the striking ability of humanity to transform the *status quo* and to begin new things, which can simply be called freedom and obviously rests upon the factual world. In this interpretation, then, in particular the transformation of nature, the individual and society would be undermined if the creative character of all these domains were denied. From this perspective, it thus seems reasonable to understand the transformative capability of humanity as dependent co-creation in a broader sense.

Firstly, and perhaps most intriguingly in Barbour's thought, nature is intelligible. He maintains that, in contrast to other ancient forms of religion, the Hebrew and later the Christian traditions (the latter being dependent on the first) considered nature neither sacrosanct (divine) nor evil (daemonic) (Barbour 1970; Barbour 2000).[6] On the contrary, the intelligibility of nature stems from nature's order, which allows for understanding, intervention, and science (Barbour 1990).[7]

Secondly, Barbour encourages an understanding of the *dominium terrae* (Gen 1:28), and the ambivalent imperative to subdue earth, as a summons to intervene – a task for which technology could be a major means. He asks, in a noticeably positive account of technology and modernity: "Can we not think of technology as a fulfillment of the commission to subdue the earth? Only in our century has such a dominion become effective – dominion over microorganisms by medical advances, dominion over space by supersonic planes and space flight, dominion over time by television communications linking the world […]" (Barbour 1970, 60).

Thirdly, Barbour cites the Dutch theologian, Arend van Leeuwen (1918–1993), who perceives the *exodus* tradition at work in the liberating potential of technology, resulting in a Christian understanding of technology as "Christ incognito": "Van Leeuwen asserts that technology is today bringing precisely the same kind of liberation to newly emerging nations, challenging the sacred order of traditional cultures, shattering their established institutions, and opening man's life to a new future unlike the past. Technology, as he says, is Christ incognito, liberating man from the authority of tradition" (Barbour 1970, 60; Van Leeuwen 1965). He finds a similar, yet more general concept in the *œuvres* of Friedrich Gogarten (1887–1967) and Dietrich Bonhoeffer (1906–1945). Both understand Christian faith as fostering and accompanying natural and societal developments. Gogarten, as Barbour summarizes, understands faith as an agency that "[…] frees man from bondage to the power of the world" (Barbour 1970, 61). Furthermore, Bonhoeffer described a process leading to "[…] maturity, autonomy, and adulthood", implying (in contrast to ages of religiously imposed immaturity) that "[m]odern man no longer needs God as […] a helper" (Barbour 1970, 61). In Barbour's reading of Gogarten

6 Working with what earth offers as commanded in biblical texts on the Sabbath (Ex 20:8–11 and Dtn 5:12–15), or in the imperative "Niru lachem nir!" (Hos 10:12 and Jer 4:3), which means "Take new land under the plow", characterizes the Jewish and Christian approach to creation in a very fundamental sense. In other words, creation is not sacrosanct nor untouchable, but material to be worked upon.

7 Countering the danger of a science-averse Christian theology, Barbour recalls the high number of Puritans in the Royal Society in London (founded in 1660, and the first institution to support advances in science) as an early example of a science-affirming Christianity (Barbour 1990).

and Bonhoeffer (which, admittedly, requires further scrutiny), with regard to the question of its role over nature, humankind seems to play an active part in nature's transient processes. Moreover, following Leeuwen, the above powers seem to be enfolded in theology to the extent that technological intervention could be interpreted as Christian discipleship in order to allow for liberation from constrictive tradition and parts of nature.

But what is Barbour's own stance on technology and its meaning for humankind's role in creation? On the way to a formulation of his own view, he differentiates between enthusiasts and critics of technology. The supporters welcome technology as a liberator and the non-supporters reject technology as an enslaver; whilst the first group states that technology improves life quality and extends life expectancy, the second group insists on the threat to individuality arising from objectifying technologies that are unable to valorize subjectivity. Barbour relates the critique of technology mainly to what he names the "existentialist tradition" and to Søren Kierkegaard, Gabriel Marcel, Paul Tillich, and Albert Camus (Barbour 1970, 61–63). On the basis of this he concludes that technology can bring both tremendous benefits and enormous risks (Barbour 1990), which leads him to conclude that technology itself is "ethically neutral" (Barbour 1970, 64). In conversation with Bonhoeffer's work on religious maturity, Barbour stresses the conviction of humanity as *cooperator Dei*, suggesting a criterion for theologically evaluating action based on his own reading of the bible and Christian tradition: "humanizing the world" (Barbour 1970, 73). Barbour highlights the active role of humanity in the present and resists the theological option of delegating profound changes to the realm of eschatology: "We need not hesitate to affirm life in the world or to celebrate the secular. […] The gospel is not the enemy of human freedom and fulfillment; it liberates us to discover our true humanity, it frees us for creativity and service" (Barbour 1970, 73). Barbour's overall positive assessment of technology becomes particularly visible when he declares technology to be the ethical response to the need of the other. His remarkably positive evaluation of technology[8] occurs in the name of care: "Our response to the neighbor in need has to be expressed technologically. […] Today it is not enough to feed the poor, for we have in our hands to abolish hunger

8 Technology must expect a certain amount of reservation when it encounters contemporary theology. This holds true for unestablished forms of technological application to human bodies, especially when not primarily done in the name of therapy as presently understood, but when bodies are modified for very personal reasons. Docetism (Saskia Wendel), Pelagianism (Magnus Striet), and "Sarcism" in the apostle Paul's sense of a sinful focus on the flesh *(σάρξ)* (Ulrich Körtner) are articulated by contemporary Christian theologians as heresy-accusations against technologically enhanced bodies (Wirth 2018b).

and poverty" (Barbour 1970, 74). Barbour views today's challenges concerning technologies and calls "humanizing of technological revolution" (Barbour 1970, 74) a possible, yet very general, response. This, again, is remarkable, as particularly in theology a more or less sublime reservation against technology persists. For it smacks of a futurity without God and humanity, as the first would not be needed anymore and the second would be overrun.

Barbour's book *Science and Secularity* (1970) was written in the context of the newly discovered DNA, and all its associated (over-selling) hopes and (over-selling) fears (Barbour 1970), which gave him the opportunity to apply and concretize his aforementioned approach to technology. Impressively, for this early state of the debate, he understands and discusses the potential of gene engineering, and most astonishingly, describes genome editing and gene surgery as future possibilities: "Deliberative alterations in genetic structure are foreseen by some geneticists. A desired DNA segment might be incorporated in a virus and used to replace a particular gene segment. Chemical or microsurgical techniques of genetic programming might allow man to be remade" (Barbour 1970, 85). He inaugurates an ethical debate on "genetic control", stating that genetic control, for the purpose of detecting and eliminating defects ("negative eugenics"), is ethically harmless. He does so without any differentiation between types of genes, something he later returns to when referring to possible long-term effects of manipulating germ lines (Barbour 2000). Barbour then opposes other theological voices that believe that genetic manipulation is an offence against the dogma of the good creation. Barbour repudiates this and stresses the ethical duty to reduce diseases – he mentions mentally ill children – and calls the natural fallacy to mind, which, in the domain of theology, tends to misunderstand the present nature as exactly mirroring God's intentions (Barbour 1970), thus restricting humanity to a deterministic self-understanding (Barbour 2000). But his essential point on the forecasted gene surgery is the *novum*-character of genetic manipulation, in view of which he stresses that established ethical theories cannot simply be rephrased. Since he is aware of the ethical questions associated with genetic surgery, he suggests a general ethical criterion that is similar to Whitehead's aforementioned aim of processes:[9] "contribution of an action to human fulfillment" (Barbour 1970, 88).[10] This implies that Barbour

9 Barbour finds that process theology valorizes creation as an open and dynamic *creatio continua* (e.g. Ps 104; Ps 14:30). Besides, the responsivity that characterizes humankind in this thinking explains much and appears to be the most promising model in theology of creation and salvation, although it also has its limitations (Barbour 1990; Barbour 2000).

10 Certainly, Barbour is far from thinking of just fulfillment and technological progress together. Particularly from a biblical perspective, which Barbour refers to throughout his

does not exclude any genetic technology *per se*. He encourages an affirmative stance on transformative action, creativity and science as a habit of theology and the church against widespread neophobia in these circles; although many of them do not hesitate to become consumers of advanced technology. Barbour speaks out against any essentialist view of creation that hinders the transition of nature and society: "I don't think we should exclude any of these genetic proposals simply because it represents a radical departure from the past. [...][Man] is free to transform mankind in unprecedented ways. He must be open to genuinely new opportunities for the emergence of higher values. Creation is not complete. We are participants in creativity, helping to shape the future. Too often in the past the church has resisted change and failed to respond to the creative activity of God in history" (Barbour 1970, 89; cf. also Polkinghorne 2007). What is also obvious here is that Barbour does not mistakenly overestimate science, nor does he expect science to solve all malign situations and constitutions. According to Barbour, as a cardinal area of human creativity, science helps build an improved future. But he does not present it as a means for messianic misunderstanding of humans' role in salvation.[11]

---

books, human fulfillment can be found in respecting the entire material and immaterial dimension of the person, not playing one off against the other: "The bible presents images of human fulfillment that do not ignore material welfare; we are called to concern for the hungry and the homeless. But it also describes others' sources of fulfillment in the interpersonal relationship, appreciation of the natural world, and spiritual growth" (Barbour 1990, 217).

11 In particular, Barbour distinguishes between "prophetic eschatology" and "apocalyptic eschatology". The prophetic version originally builds the eschatological future on the moral behavior of Israel. He recalls the idea that when the nation changes its habits, then the ἔσχατον can be realized (Barbour 1990). Barbour modified this in dialogue with newer theology and understands "prophetic eschatology" as the concept of a future to come, built on a "combination of divine initiative and human response" (Barbour 1990). In this view, his concept of a humanity active in profoundly changing nature finds eschatological plausibility without imposing entire soteriological responsibility on humanity. "Apocalyptic eschatology" rejects humanity's ability to perform any action that can be considered part of the eschatological future. Here it is the supernatural alone that intervenes and causes a new future (Barbour 1990). However, while both eschatological viewpoints adequately focus on certain aspects, neither can capture the whole picture. The notion of fundamental passivity *(mere passive)* stresses the final responsibility of God in the salvation process, enabling human action as liberated from extreme future pressure. In other words, there are aspects in which humanity participates in the "eschatological now", but there are other aspects of an "eschatological not yet", for instance the impossibility *ex parte hominis* to forgive the unforgivable, especially in the name of victims who cannot forgive anymore, because they have been slaughtered (Wirth 2016).

Beyond the theological-ethical criterion for genetic manipulation (its impact on human fulfillment, which obviously involves much more than solely therapeutic aims [Barbour 1990]), Barbour concludes his chapter on the "biochemical man" by drawing attention to the issue of the possible side-effects of invasive genetic modification. Although his plea for "great caution" seems commonplace, it presents two relevant concerns: that human creativity not be hindered, and that there be transitions to a fulfilled future whilst not allowing for blind interventions and actions that can easily be regretted (Barbour 1970). Although Barbour's work does not deliver any elaborate ethical theory, he cites human fulfillment and caution as topics for theological ethics, which, obviously needing more detail and application, nevertheless serve as important, theologically insightful perspectives.

4 Arthur Peacocke and God as Encouraging Science and Technology

The British biochemist and theologian Arthur Peacocke (1924–2006) emphasizes that scientific developments, especially predicting and controlling of natural processes, have a tremendous impact on the theological understanding of God and God's relationship to nature and humanity (Peacocke 2001). The intelligibility (i. e. prediction) and transformability (i. e. controlling) of nature became a view that forced theology to reconsider assumptions about a solely active God and a solely receptive humanity and nature. Because truth and seeking for it appears as vital part of being a created creature and because this truth became the focus of theology, and not "an authoritative book" (since the bible cannot exist as a heteronomous collection), nor "an authoritative community" (since the churches cannot exist as communities of heteronomous beliefs and limited knowledge), new scientific discoveries must have a deep influence in terms of shedding light on the truth about creation (Peacocke 2001). One argument Peacocke presents in order to show how scientific research relates to theological reflection about truth is that the discovery of (biological) diversity can be interpreted as mirroring God's intention (Peacocke 2001).

What might prove important for the theological and ethical evaluation of genome editing in conversation with Peacocke is his note on the developmental characteristic of Christian doctrine and the metaphors that explain that doctrine in dialogue with scientific developments. He stresses that this is typically overlooked by Christians, who tend to think of doctrine as being rather solid (Peacocke 2001). In this vein, genome editing, as a new mode of transforming nature and creation, shows the profound modifying potential of nature in a new and different way. Also,

Peacocke assumes that a positive reception of "new understandings of nature" (the insights and methods of genome editing are just one example) could support the credibility of religion in its confrontation with science (Peacocke 2001). As it turns out, "explorations towards God can be based only on [...] reflection on nature and humanity" (Peacocke 2001, 16), which follows the conviction that the first word of God appears within nature and humanity (Peacocke 2001). Nothing seems more accessible for the study of God than creation itself, although other paths, like intellectual speculation, mysticism, etc., may also allow for approximations (Peacocke 2001).

One prime example of discovering God in creation is the observation that persons are active, creative, and transformative entities. In theological terms, humans appear as "creating creatures", and due to their factual realm of possibilities, they cannot be understood as "mere playthings of God" (Peacocke 2001, 86). Peacocke asserts that, along with the desire for "the existence of self-conscious, intelligent, freely willing persons", comes a new understanding of God and creation. Neither can be interpreted as closed realities when persons are involved in real, and not only fictive, interaction. According to Peacocke, both God and creation are receptive, which implies the possibility of God's suffering and creation's transformation (the "existence of new and hazardous possibilities") (Peacocke 2001, 89). Encapsulated within this lies the risky element of an open composition of creation. But, in line with the "free will defense" (Polkinghorne 2007, 71), Peacocke deems side effects a necessary cost for a God who created a "freely responsive humanity" (Peacocke 2001), in which only free persons are able to become co-lovers *(condilecti)*. By Peacocke's line of reasoning, however, God does not appear to be imprisoned in the mode of self-limitation, but communicates with humanity, always mediated through religious experiences, the experience of significance, etc. (Peacocke 2001). Peacocke then interprets this to the effect that "God's interaction with and influence on the world and its events" (Peacocke 2001, 113) can affect both the mental and biological configuration of a person. Obviously, human beings are objects of effects from "the non-human world". Yet, following Peacocke, "human agency in the non-human world" is also observable (Peacocke 2001, 113) and poses a problem to the understanding of humanity as being merely passive. Peacocke hypothesized that this is what could be understood as *creatio continua*, which "is the immanent Creator creating continuously in and through the processes of natural order" (Peacocke 2001, 129). Notably, Peacocke favors an interactive concept of a *creatio continua*. This means that, in his thinking, God's power does more than only maintaining creation as a whole, without influencing concrete developments and patterns. Yet this is exactly what Peacocke considers to be the profile of continuous creation, so that through "mediated influences" concrete impact is possible (Peacocke 2001).

With this comes the claim of the immanence of God, who, through the lens of Peacocke's thinking, does not act through "supposed gaps" within natural law, and thus, does not appear as a wondrous "additional factor", but "is 'in, with and under' all-that-is and all-that-goes-on" (Peacocke 2001, 146; cf. Predel 1996; Smedes 2004).

It is not necessary to point out the multitude of concerns raised by a stance with monistic, pantheistic and instrumental components in its understanding of the God-world-relation, especially where the question of theodicy is concerned (Peacocke 1990; cf. Smedes 2004).[12] But such a position at least draws attention to the theological interest in valorizing new developments in science as potentially congruent with a Christian worldview, and offers more than theology's notorious bio-conservative reflex. Peacock's plea for "a more dynamic view on God's continuous action in the processes of the natural world" (Peacocke 2001, 132; Peacocke 1990) needs to be confronted with a critique that questions the factuality and possibility of concrete divine action within natural processes. His theory is rather ignorant of the malignant parts of nature that still cause severe sickness and premature death. Thus, his theology verges on either naïve fideism (belief without grounding in reason) or the presentation of a cruel and quite unsuccessful God.

This potentially shattering critique aside, those who still wish to learn from Peacocke could conclude, in the debate about invasive interventions into nature (as is the case with genome editing, for instance), that new technological possibilities are capable of mirroring God's mediated will for the further improvement of his creation and God's aversion to sickness.[13] But this raises an important question: Why not see these developments as the fruits of both the intelligibility of nature and the intelligence and experiences of humanity? Why must God be actively involved? Peacocke, however, stands by the liberal theology characterized by the expectation that God causes new meaning through scientific developments (Peacocke 2001). He thus advocates understanding classically secular areas as religiously laden, welcoming new developments in science as the possible result of divine-human-interaction. His account of God's developing interaction with persons and nature implies process thinking, which becomes obvious when he addresses Jesus as a "learner"

12 Peacocke does not overlook the evidence of malignancy in creation, and he admits that his concept does not deliver any approximate solution, other than referring to the believed-in divine self-limitation. In this area, Peacock supposes potential for dealing with theodicy (Peacocke 1990).

13 John Polkinghorne puts it similarly with a very optimistic outlook on history, interpreting it as "grand improvisation": "[...] theology does not need to see the history of the world as the performance of a fixed score, written by God from all eternity, but may properly understand it as the unfolding of a grand improvisation in which the Creator and creatures both participate" (Polkinghorne 2004, 67–68).

or "house-holder" who can create new things from existing elements (Peacocke 2001, 134). But all divine interaction with nature and persons, in a mediated way, does not cause extensive fulfillment. His *creatio continua* is obviously faced with obstacles and requires the (eschatological) transformation (but no re-creation in the sense of a *creatio ex nihilo*, but a *creatio ex vetere* [Polkinghorne 1998]) that, in Peacocke's concept, begins with a very close linking of God, persons and nature (Peacocke 2001).

5 Conclusion

Both process thinking and related concepts of a dynamic creation in theology rest upon the idea of a *contraction Dei*, a self-limitation of God for the sake of the other (Barbour 1990; Peacocke 2001; Polkinghorne 1998; cf. also Clayton 2004; Dabrock 2009; Maaßen 1990; Predel 1996; Smedes 2004). The idea of self-limitation or contraction implies a divinity who, by limiting its own power, allows for the existence of creation, freedom and otherness. Process thinking, implying the *contraction Dei*, is not without limitations. Yet it allows for creative thought, freedom of humankind and a good end guaranteed by God (Rev 21:1) (Müller 2009; Welker 1988; Wirth 2016). Where the creature is thought of dynamically, divine withdrawal is not a product of God's indifference but instead acts as a means of enabling freedom that is naturally involved in a process of positive and negative developments. This freedom is constitutively capable of being addressed – Whitehead gave distinction to the idea of persuasion defined love as God's *modus operandi* (Maaßen 1990; Müller 2009; Sander 1991) – and, if it were untrue that God invites limited freedom and limited processes to receive eschatological transformation (Rev 21:5 "See, I am making all things new"), human intervention would be overburdened. Since, in process thinking, God offers humanity the maximum room for freedom of development and selfhood (Welker 1988), there is no contradiction of the implicit withdrawal when God eschatologically transforms the beginnings of fulfillment, achieved by human activity, into the mode of definite fulfillment (Whitehead 1977). This is because the scope of divine contraction remains the same but acknowledges different stages in the allowance of development and selfhood. Withdrawal does not literally mean to vanish completely, but allows for certain kinds of intervention, especially for the sake of the otherness of the other, for which the contraction occurs. From the perspective of dynamic concepts of creation, accusing God of all the miserable developments in the processes of the individual and of world history would mean undermining both the capability of humanity for improvement (for instance through science [Hampe 1998]) and the necessary open space for becoming a self. The fact that there are no limitations, especially where evil is concerned (Peacocke 2001), leads to the general question of theodicy. But this is not a conundrum that only process theology faces.

In any case, divine contraction appears as a *conditio sine qua non* for any creative and free action of humans, in the private as well as the scientific domains. After the elaboration of concepts of process theology, and a dynamic understanding of creation and nature, it is open to discussion whether or not the technologies of genome editing for enhancement purposes could be understood as a way of transforming nature in the name of the creator, who obviously allows for interventions into the intelligible creation. In the scenario that genome editing procedures lead to the next steps of human fulfillment, genome editing could be considered in accordance with the aims of the creator.

Theologically speaking, invasively transforming the natural base of the human does not alone appear to be problematic, when considered from the perspective of process theology and dynamic conceptions of creation. For an ethical critique of a profound transition of genetic domains, more substantial arguments must be made than the simple, dismissive rebuttal that genome editing, for instance, constitutes an impermissible interference in God's good creation. Obviously, many external factors have an enormous impact on our biological constitution.[14] The food we eat, the air we breathe, and the environment we live in influence our biology irreversibly; all of this appears as something external outside but with time it is integrated into the body (*integrierte Gesellschaften*) (Hampe 1990, 228). And, over time, this can impact the future of possible offspring in positive and negative ways.

It is undeniable that there are debatable ethical issues that come with genome editing, although this was not the focus of this paper. Nevertheless, these shall be mentioned briefly, in order not to underestimate the problematic nature of genome editing. The central difficulty with genome editing is not its potential to transform human nature, but possible forms of repression, for example concerning normative corporeality or economic exclusion. Moreover, there is the danger of a solely objectifying approach to human beings in the application of these and other technologies. These are the questions that should occupy theological ethics, rather

14 Michael Hampe underlines the connection between an organism and the environment in his study on Whitehead. Through the constant exchange of material, the outside becomes part of the inside of the organism in the very fundamental sense of being part of the causal connection of an organism: "Der ständige Materialaustausch in einem Lebewesen macht es nötig, dass es durch Atmung und Ernährung ununterbrochen Gesellschaften aus seiner Umwelt in den Kausalzusammenhang des Organismus integriert" (Hampe 1990, 228).

than the perpetuation of worn out and accusatory religious convictions about an untouchable creation.

Since the aim of this study was to draw theological attention to the ways in which genome editing could contribute to the next steps towards human fulfillment (in a reality characterized by processes), but also towards possible harm and repression, the title of this paper reveals its double meaning: transition and care stands for "transition in the name of care", but it also means "care for transition". Both elements are required for a balanced position on genome editing and other technological modes of human alteration (Dabrock et al. 2010). In particular, the merits of process thinking in theology, including all limitations and weaknesses, help to avoid the blindness of both over-estimation ("hype") and under-estimation (fear).

References

Baltimore, David, et al. 2015. A Prudent Path Forward for Genomic Engineering and Germline Gene Modification. *Science* 348: 36–38.

Barbour, Ian G. 1970. *Science and Secularity. The Ethics of Technology.* New York/Evanston/London: Harper & Row.

Barbour, Ian G. 1990. *Religion in an Age of Science. The Gifford Lectures 1989–1991. Volume 1.* London: SCM Press.

Barbour, Ian G. 2000. *When Science Meets Religion. Enemies, Strangers or Partners?* San Francisco: Harper.

Case-Winters, Anne. 2004. System and Dynamism in Whitehead's Thought: The Category of the Ultimate and the Concept of God. In *Schleiermacher and Whitehead. Open Systems in Dialogue*, Helmer, Christine (ed.), 135–158. Berlin/New York: de Gruyter.

Clayton, Philip. 2004. Systematizing Agency: Toward a Panentheistic-Participatory Theory of Agency. In *Schleiermacher and Whitehead. Open Systems in Dialogue*, Helmer, Christine (ed.), 215–233. Berlin/New York: de Gruyter.

Dabrock, Peter. 2009. Playing God? Synthetic Biology as a Theological and Ethical Challenge. *Systems and Synthetic Biology* 3: 47–54.

Dabrock, Peter, and Jens Ried. 2010. Leben machen – Gott spielen? Theologische Anmerkungen zur Ethik der Synthetischen Biologie. *Die Politische Meinung* 55: 37–41.

Dinter, Astrid. 1999. *Vom Glauben eines Physikers. John Polkinghornes Beitrag zum Dialog zwischen Theologie und Naturwissenschaften.* Mainz: Matthias-Grünewald-Verlag.

Fetz, Reto Luzius. 1990. Whiteheads Begriff einer Religion im Werden und die Theorie der Moderne. Eine Interpretation Whiteheadscher Religionstheorie im Kontext gegenwärtiger Sozialwissenschaft. In *Natur, Subjektivität, Gott. Zur Prozeßphilosophie A. N. Whiteheads*, Holzhey, Helmut, Alois Rust, and Reiner Wiehl (eds.), 278–300. Frankfurt: Suhrkamp.

Griffin, David Ray. 2004. Feeling and Morality in Whitehead's System. In *Schleiermacher and Whitehead. Open Systems in Dialogue*, Helmer, Christine (ed.), 265–294. Berlin/New York: de Gruyter.
Hampe, Michael. 1990. *Die Wahrnehmung der Organismen. Über die Voraussetzungen einer naturalistischen Theorie der Erfahrung in der Metaphysik Whiteheads.* Göttingen: Vandenhoeck & Ruprecht.
Hampe, Michael. 1998. *Alfred North Whitehead.* München: C.H. Beck.
Kipke, Roland, Markus Rothhaar, Martin Hähnel 2017. Contra: Soll das sogenannte „Gene Editing" mittels CRISPR/Cas9-Technologie an menschlichen Embryonen erforscht werden? *Ethik in der Medizin* 29: 249–252.
Koslowski, Peter. 2002. Fortschritt, Apokalyptik und Vollendung der Geschichte und Weiterleben des Menschen nach dem Tod in den Weltreligionen. Einleitung. In *Fortschritt, Apokalyptik und Vollendung der Geschichte und Weiterleben des Menschen nach dem Tod in den Weltreligionen*, Koslowski, Peter (ed.), 1–9. Paderborn: Wilhelm Fink.
Lamm, Julia A. 2004. The Force of Dialogue and a Dialogue of Forces: Resources for Open Theological Systems. In *Schleiermacher and Whitehead. Open Systems in Dialogue*, Helmer, Christine (ed.), 237–263. Berlin/New York: de Gruyter.
Leeuwen, Arend van. 1965. *Christianity in World History.* New York: Charles Scribner's Sons.
Maaßen, Helmut. 1990. Gottes Beziehung zum Guten und Bösen in Whiteheads relationaler Wertethik. In *Natur, Subjektivität, Gott. Zur Prozeßphilosophie A. N. Whiteheads*, Holzhey, Helmut, et al. (eds.), 262–277. Frankfurt: Suhrkamp.
Müller, Tobias. 2009. *Gott – Welt – Kreativität. Eine Analyse der Philosophie A. N. Whiteheads.* Paderborn: Schöningh.
Peacocke, Arthur. 1990. *Theology for a Scientific Age.* Oxford: Blackwell.
Peacocke, Arthur. 2001. *Paths from Science Towards God. The End of All Our Exploring.* Oxford: Oneworld.
Polkinghorne, John. 1996. *Beyond Science. The Wider Human Context.* Cambridge: Cambridge University Press.
Polkinghorne, John. 1998. *Belief in God in an Age of Science.* New Haven: Yale University Press.
Polkinghorne, John. 2000. Eschatology: Some Questions and Some Insights from Science. In *The End of the World and the Ends of God. Science and Theology on Eschatology*, Polkinghorne, John, and Michael Welker (eds.), 29–41. Harrisburg: Trinity Press International.
Polkinghorne, John. 2004. *Science and the Trinity. The Christian Encounter with Reality.* New Haven/London: Yale University Press.
Polkinghorne, John. 2007. *Quantum Physics and Theology. An Unexpected Kinship.* New Haven/London: Yale University Press.
Predel, Gregor. 1996. *Sakrament der Gegenwart Gottes. Theologie und Natur im Zeitalter der Naturwissenschaften.* Freiburg: Herder.
Rohmer, Stascha. 2000. *Whiteheads Synthese von Kreativit*ät und Rationalität. Reflexion und Transformation i*n Alfred North Whiteheads Philosophie der Natur.* Freiburg/München: Karl Alber.
Sander, Hans-Joachim. 1991. *Natur und Schöpfung – die Realität im Prozess.* Frankfurt am Main: Peter Lang.
Schramm, Michael. 1991. *Prozeßtheologie und Bioethik. Reproduktionsmedizin und Gentechnik im Lichte der Philosophie A. N. Whiteheads.* Fribourg: Universitätsverlag Freiburg.
Smedes, Taede A. 2004. *Chaos, Complexity, and God. Divine Action and Scientism.* Leuven: Peters.

Whitehead, Alfred North. 1969. *Adventures of Ideas*. New York: Macmillian.
Whitehead, Alfred North. 1977. *Process and Reality. An Essay in Cosmology*. New York: The Free Press.
Whitehead, Alfred North. 1990. *Science and the Modern World*. New York: The Free Press.
Welker, Michael. 1988. *Universalität Gottes und Relativität der Welt. Theologische Kosmologie im Dialog mit dem amerikanischen Prozessdenken nach Whitehead*. Neukirchen-Vluyn: Neukirchner Verlag.
Wirth, Mathias. 2016. Trostlose Eschatologie? Zu einer unerledigten Kontroverse über Versöhnung in der neueren Dogmatik. *Neue Zeitschrift für Systematische Theologie und Religionsphilosophie* 58: 259–284.
Wirth, Mathias. 2018a. Trans-Körper. Theologie im Gespräch mit Transhumanismus und Transsexualität. *Zeitschrift für Evangelische Ethik* 62: 10–30.
Wirth, Mathias. 2018b. Doketisch, pelagianisch, sarkisch? Transhumanismus und technologische Modifikationen des Körpers in einer theologischen Perspektive. *Neue Zeitschrift für Systematische Theologie und Religionsphilosophie* 60: 142–167.

Human Germline Genome Editing in the Clinical Context

The Case of Disease Prevention

Giovanni Rubeis

Keywords

Assisted Reproduction Technologies, Bioethics, CRISPR-Cas9, Germline, Reproductive Autonomy, Wish-Fulfilling Medicine.

Abstract

Interventions in the human germline have been regarded as a red line in genetic engineering up to now. However, with the recent progress in genome editing techniques, first and foremost CRISPR-based methods, the tide seems to be turning. The therapeutic benefits in particular are brought forward as an argument in favor of germline genome editing. According to this view, the main benefit of GGE is disease prevention. The procedure could be used to prevent monogenetic conditions for which no treatment is available so far. This leads many commentators to the view that the benefits of GGE outweigh the risks. In this article, I examine the status and ethical implications of GGE as a means of disease prevention. I intend to show that the notion of GGE as disease prevention is only true for one specific application. Instead of being preventive treatments, most applications of GGE are advanced techniques of assisted reproduction. As such, they can be thought of as wish-fulfilling medicine, that is the use of medical techniques for non-medically indicated purposes. Thus, it is necessary to reconsider the risk-benefit analysis of GGE for most of its applications.

M. Braun et al. (Eds.), *Between Moral Hazard and Legal Uncertainty*, Technikzukünfte, Wissenschaft und Gesellschaft / Futures of Technology, Science and Society, https://doi.org/10.1007/978-3-658-22660-2_10

1 Introduction

Interventions in the human germline for clinical purposes have been discussed since the 1980s. Since then, severe medical risks as well as ethical considerations have been put forward against germline interventions. From a biomedical point of view, it is simply not possible to predict the exact outcomes of modifying the germline for an individual, much less for coming generations. Also, the available methods have long been insufficient. A whole bundle of concerns has been voiced from an ethical point of view. Some have seen germline intervention as a red line, in the interference with nature, that must not be crossed. Others have feared societal implications like a new eugenics. However, this attitude has been challenged in recent years, especially due to astonishing progress in the development of genome editing methods. The decisions of licensing bodies and other institutions played an important role in this respect. In 2016, the UK Human Fertilisation and Embryology Authority (HFEA) licensed germline interventions for basic research using the CRISPR-Cas9 method. The research project at the Francis Crick Institute in London aims at gaining new knowledge about early embryonal development, the differentiation of stem cells, and the mechanisms behind infertility. Kathy Niakan and her research team use supernumerary zygotes (the pre-embryonic state of fertilized oocytes) from in vitro fertilization (IVF). The zygotes are developed until they reach the blastocyst state and include up to 256 cells. After seven days, these blastocysts are destroyed. As a first important result the team managed to block the OCT4-gene (Fogarty et al. 2017), which triggers embryonal development. Although this research project is labelled as basic research, and the blastocysts are not transferred in utero, it is the first project that involves the genetic editing of the human germline. In 2017, the door for clinical trials using genome editing methods was opened for the first time. In their report *Human Genome Editing: Science, Ethics, and Governance*, the American National Academies of Science, Engineering, and Medicine stated that clinical research using germline genome editing (hereinafter: GGE) in humans should be permitted under certain specific conditions (The National Academies 2017). The National Academies propose to limit GGE to severe cases of disease and disability where no alternative treatment is possible. In the long run, this could lead to the development of clinical applications.

It is usually agreed upon that the main clinical perspective of editing the human germline is disease prevention (Ishii 2015; Long et al. 2014). The CRISPR-Cas9 method in particular could prevent severe monogenetic conditions, like Huntington's disease or sickle-cell anemia, saving the lives of those afflicted and providing them with a quality of life they could never have achieved otherwise. Thus, disease prevention is widely seen as a benefit that outweighs the severe risks associated with GGE. Some even claim that the therapeutic use of GGE is a moral imperative. Following this view, we have an obligation towards individual patients and future generations to apply GGE for disease prevention (Gyngell et al. 2016). However, others claim that the medical as well as societal risks (like the use of GGE for genetic enhancement and the possible social divide between the enhanced and the non-enhanced that may follow) are too high, with some even arguing for a full ban (Lanphier et al. 2015). Since research efforts are speeding up, and clinical applications of GGE might be available in the not so distant future, it is important to have an intense ethical debate at this early stage. We are standing on the threshold of the clinical application of GGE, which demands an ethical framing, especially as a guideline for clinicians. It is therefore crucial to evaluate the status of GGE, as a means of disease prevention, from an ethical point of view.

As a first step, I closely examine the method's clinical perspectives. This is followed by an ethical evaluation that focusses on the question whether the therapeutic benefits of GGE really outweigh the risks that are usually ascribed to a germline intervention. In my analysis, I use the method of pragmatic evaluation that was established as a counterpart to categorical arguments in the context of gene therapy (Bayertz 1991). This method demands that, instead of categorically assessing a procedure of gene therapy as ethically acceptable or unacceptable, the concrete applications and their contexts be evaluated specifically by weighing their respective benefits against their risks (Rubeis et al. 2016).

2 Procedures and Therapeutic Perspectives of Germline Genome Editing

Genome editing, especially CRISPR-based methods, offers new possibilities in germline therapy (Ishii 2015; 2017). During an IVF, or an intra-cytoplasmic sperm injection (ICSI)-procedure, an editing system (programmable nucleases) could be microinjected in the zygote. Afterward, the developing embryo would be tested by using prenatal genetic diagnosis (PGD) to detect on-target-mutations as well as off-target-effects. The PGD would include a blastomere biopsy three days post fertilization, or a trophectoderm biopsy four to five days post fertilization. The resulting verified embryo would then be transferred in utero. Another possible application is oocyte editing. Oocytes would be retrieved and microinjected with an editing system. The edited oocytes would be used for IVF or ICSI. The developing embryo would undergo PGD before being transferred in utero. In spermatogonial stem cell (SSC) editing, SSCs could be retrieved (via testicular biopsy), sorted and purified. In a further step, the SSCs would be transfected with an editing system and afterwards undergo genetic analysis. The verified SSCs would be transferred back into the testes in order to produce functioning sperm cells. These sperm cells could be used for IVF or ICSI, resulting in an embryo that undergoes PGD and is transferred in utero. Another option would be in vitro gametogenesis (IVG). IVG includes the reprogramming of somatic cells into induced pluripotent stem cells (iPSCs), from which germ cells (sperms and oocytes) are derived. These germ cells could be used in IVF or ICSI, thus providing an advanced infertility treatment. Further possible applications of IVG include embryonic research, or the creation and storage of embryos for possible stem cell therapy later on. Finally, GGE might be used for correcting mutations in the mitochondrial DNA (mtDNA), thus minimizing the risk of inheriting severe health conditions.

These procedures offer new therapeutic perspectives. The perspective mainly discussed is the prevention of monogenetic diseases like Huntington's disease or beta thalassemia. On an ontogenetic level, diseases in the individual resulting from the treated embryo could be prevented. On a phylogenetic level, the spread of certain gene types within a population could be prevented. Another perspective is infertility treatment, as with the aforementioned SSC editing, or other forms of assisted reproduction, like deriving gametes from iPSCs using GGE.

These therapeutic applications are prospective; their feasibility is hypothetical, based on experimentation on cell and animal models as well as the very limited empirical evidence from recent research on human embryos (Kang et al. 2016; Liang et al. 2015; Ma et al. 2017). As mentioned above, the trial currently in progress at the Francis Crick Institute in London is also expected to yield further results in this context (Callaway 2016). Yet, the discussion about risks and benefits of germline interventions is not speculative or empirically unfounded. There have already been cases of genetic germline modification for clinical purposes, although not involving genome editing methods. In the 1990s, ooplasmic transfer was used as an infertility treatment. The method included inserting cytoplasm from a donated oocyte into an oocyte with low fertility (Gómez-Tatay et al. 2017). Although successful at first, ooplasmic transfer resulted in miscarriages, embryonal and fetal abnormalities as well as health issues of born children. The method was banned by the US American Food and Drug Administration (FDA) in 2001. Mitochondrial replacement techniques (MRT) can also be regarded as intervention in the germline, although this categorization is hotly debated (Baylis 2017). The first live birth after using MRT was reported in 2016, and so far, there seem to be no complications (Zhang et al. 2016). In all of these cases, the proponents justified the risks with the expected benefits. Ooplasmic transfer was regarded as new prospect for infertile women. MRT is seen as a method of preventing mitochondria-related diseases, allowing women who carry pathogenic mutations in their mtDNA to have children free of these mutations. The same holds for GGE, which is mainly propagated as a method of disease prevention.

3 Risks of Germline Genome Editing

The first studies using GGE on human embryos show, that although CRISPR-Cas9 means an immense improvement, there are still severe medical risks (Kang et al. 2016; Liang et al. 2015; Ma et al. 2017). For one, although CRIPSR-based methods in particular have been refined within the last years, off-target effects are still prevalent. That means that double-strand DNA breaks occur at the wrong target site, leading to an inaccurate or incomplete editing. The consequences are improper translocations, inversions, point mutations, or large deletions (Ishii 2017). Although a lot of effort is put into the refinement of guiding molecules, such as guide RNA in CRISPR-Cas9, a substantial risk will remain. Also, there is the risk of genetic mosaicism, meaning the coexistence of edited cells and wild types within the organism, which could have unforeseeable consequences. Genetic mosaicism may result from meiotic or mitotic error, leading to chromosomal imbalances. Since the germline is modified, these negative effects could be passed on to subsequent generations. This means that a risk assessment also has to include future individuals who may inherit a modified trait. As the current research shows, GGE is too unsafe for any clinical use at the moment. However, since more knowledge about the method and its effects is needed in order to reduce the risks, germline research using genome editing is considered as necessary by many (Ishii 2017; Ma et al. 2017; The National Academies 2017). For example, the National Academies claim in their report that the translational research from bench to bedside needs refined research protocols, transparent patient information and close monitoring of mid-term and long-term effects (The National Academies 2017).

Apart from medical risks, societal risk can be identified. By determining a future individual's genetic makeup, living individuals would decide over that future individual's preferences, compromising his or her right to an open future. According to the slippery slope argument, GGE would inevitably lead to non-medically indicated uses, like genetic enhancement. In this context, questions of social justice would arise. What would a society look like in which enhanced humans and wild types coexisted? It is easy to imagine how differences in genetic makeup would translate into differences in social status. Yet, although enhancement is clearly a risk, the slippery-slope argument is flawed. Enhancement with all its possible implications and issues is not a necessary consequence of therapeutic GGE. Similar concerns were voiced with the introduction of PGD, arguing that it would inevitably lead to a new eugenics. The experience of PGD shows that it is possible to limit a method to its medically indicated use through weighed policy-making and legal regulations. There is no reason why this should not also work for GGE. The possibility of enhancement has to be taken seriously, but it is not in itself a strong argument against GGE. The medical risks, however, call for convincing arguments, in the form of benefits, in order to tip the scales in favor of GGE.

4 Benefits of Germline Genome Editing

Many point to the possibility of preventing severe genetic diseases, like Huntington's, in future individual(s) as an argument in favor of GGE (Gyngell et al. 2016; Powell 2015). Reproduction partners[1] with a known genetic risk could use GGE in order to avoid passing on pathogenic traits to their offspring. The benefits of this are manifold. First, there is the benefit on the ontogenetic level. Editing the gene sequence responsible for Huntington's disease in the embryo would mean that the resulting individual will not be afflicted by that disease. Furthermore, this individual would not be a carrier of the genetic defect and would not pass it on to future generations. Thus, there is a benefit on a phylogenetic level through the removal of pathogenic mutations from the gene pool. GGE, its supporters claim, could also provide reproduction partners with a known genetic risk with the only possibility of having healthy, genetically related offspring. Finally, supporters state that GGE has an advantage over existing methods for disease prevention, like PGD, which would mean a benefit in ethical regards. This is because PGD entails selecting a healthy embryo and discarding others, while GGE cures an embryo with a genetic defect. The selecting and discarding of embryos, seen by many as an ethical issue, could be avoided by using GGE.

1 I use the term "reproduction partners" to signify the individuals in the reproduction process who contribute gametes or cellular material, women who carry the child, and persons who assume the status of legal guardian to the child. By using this term, I wish to avoid using "parents", a term that is normatively charged, and in some contexts (e.g. third-person reproduction), misleading.

5 Ethical Implications

The risk-benefit analysis seems to suggest, that although there may be certain risks, in the end, they are outweighed by the benefits of GGE. Some claim that, as a therapeutic procedure, it could save lives, making certain risks acceptable (Gyngell et al. 2017). Some commentators argue that disease prevention, be it on an individual or a public health-level (eliminating genetic defects from the gene pool), needs to be seen as a responsibility (Powell 2015). According to these views, the beneficence principle, the doctoral mandate, and public health-considerations make the use of GGE for disease prevention a moral obligation (Gyngell et al. 2017; Harris 2016; Powell 2015). The procedure should be used to prevent conditions for which no treatment is available, for example, Huntington's disease. Following this approach, the advantage of GGE over alternative methods like PGD is that, instead of selecting one embryo and discarding others, it is used to prevent disease in one particular embryo. Thus, the ethical issue of selection could be circumvented. According to its supporters, another important argument in favor of GGE is reproductive autonomy. Denying access to a method that has proven efficient would be hard to justify in a liberal democratic society.

At first glance, the benefits of GGE seem to outweigh the risks under certain conditions. There seems to be no reason to oppose the clinical use of GGE, provided that research efforts are intensified, the safety issues are tackled, and the methods are further refined. Even if one doesn't maintain that there is a moral obligation to use GGE, there still appear to be strong arguments in favor of the procedure. These arguments are based on the assumption that GGE is a therapeutic method, a means of disease prevention. But this assumption is false. When we scrutinize the methods mentioned above, we find that except for zygote editing, GGE does not prevent an individual from having a disease. Rather, it creates a healthy individual instead. This is what is usually called the non-identity problem. Instead of curing individual A of a disease (modifying it so it becomes A'), another individual, B, is created. A and B are not identical. They are not the same individual only with altered genes, but two different individuals. In other words, GGE does not treat, cure, or prevent disease in an individual. When it comes to zygote editing, the concept of disease prevention is applicable. However, the edited zygote would not have been produced at all if it were not for the method of GGE. The existence of the method alone leads the reproduction partners to create a zygote that is immediately edited, producing an embryo without the pathogenic genetic trait. The use of the term "therapy" is simply a misnomer here. There was no individual who was worse off before the application of the method and is better off now. Except for zygote editing, GGE is key in creating a healthy embryo, not in curing a diseased one. Therefore, GGE can only be regarded as a therapy for disease prevention in some cases and with reservation. Considering most applications, GGE is first and foremost a method of assisted reproduction that can be used for creating healthy offspring.

As an assisted reproduction technique, GGE has to be compared with existing methods. Some authors claim that GGE is the better alternative to PGD, because it avoids the issue of selecting and discarding embryos (Gyngell et al. 2017). This is not the case at the moment, at least not completely. For, according to the current state of knowledge, GGE has to be combined with PGD to investigate the desired modifications and possible off-target effects. That means there is no medical possibility to use GGE instead of PGD. However, the use of GGE could reduce the quantity of discarded embryos which would be an advantage. Some claim that GGE might be a viable method for conditions where PGD cannot be applied (Gyngell et al. 2016; Ishii 2017). This is the case with autosomal dominant diseases with at least one homozygous reproduction partner. An example would be Huntington's disease. Furthermore, PGD is not applicable for autosomal recessive diseases where both reproduction partners are homozygous, which is the case with cystic fibrosis. However, some form of PGD will still be necessary to check whether the intended editing has worked. Whether GGE really is a viable alternative to existing methods will depend on the further refinement of the method. If GGE reduced the number of embryos needed, thus also reducing the IVF-cycles the woman has to go through, this would surely be an advantage over existing methods of assisted reproduction.

That GGE is not a means of disease prevention is not of purely academic interest. Rather, the foregoing reevaluation of the method also has ethical implications, which have consequences for clinical practice. When it comes to risk assessment, it makes a difference whether we are talking about a valuable therapeutic method to prevent disease, or an elaborate assisted reproduction technique to which there are existing alternatives. In the first case, we would be willing to take risks in some cases, as long as they are outweighed by the benefits. This is exactly the argument the National Academies used when they suggested translational research for a future clinical application of GGE (The National Academies 2017). The method, the National Academies claim, should be restricted to a preventive use in cases of severe disease or disability where no alternative methods exist. This is a reasonable way to deal with a new and risky technology. However, both criteria are not fulfilled with all applications of GGE. Except for zygote editing, GGE is neither a method of disease prevention, nor is it without alternatives. First and foremost, GGE could be used as an assisted reproduction technology in cases where there is a certain risk of transmitting a genetic disease. In other cases, GGE procedures like, for example, SSC-editing, may be the only option for reproduction partners to have offspring to whom both are genetically related. In the context of assisted reproduction and infertility treatment, GGE may be a real alternative to existing methods. Reproduction partners could also use gamete donation, or even adoption, as a non-medical option to fulfil their wish to have a child. However, genetic relatedness to both reproduction partners would not be possible. The importance of genetic relatedness should not be underestimated. However, it is highly debatable whether the desire to have a genetically related child justifies the use of GGE. First, the risk for the child has to be taken into consideration. In each individual case, the risk assessment would have to show that GGE is the best option. Second, there is the issue of possible consequences for future generations. The decision of the reproduction partners to use GGE would not only affect one individual, but possibly many generations to come. This means that we have to weigh the reproductive autonomy of these individuals against the safety of many future individuals. There is also a conflict involved between the autonomy of living individuals and the autonomy of future individuals. These future individuals would be affected by the possible negative consequences of GGE without having had the possibility to consent to the method. The question is whether the desire to have genetically related children is sufficient to justify the potential risks for future generations. Finally, there are societal costs connected to the further development, implementation, and application of GGE. Financial resources as well as manpower and equipment are needed. The rhetoric of a life-saving therapy, used by the proponents of GGE, aims at justifying these costs. Who could reasonably oppose the view that we should invest resources in technologies that may save lives? As long as GGE is regarded as a therapeutic method for disease prevention, it could also be argued that collective funding of its clinical application is justified as a part of the universally funded public health service. However, things change when we regard GGE as what it mainly is: an assisted reproduction technique from which only a small number of individuals benefit. As such, GGE does not constitute a medical necessity; it is a means of fulfilling the desire to have a healthy child that is genetically related to both reproduction partners. Given this change of perspective, it is not to be expected that the public would provide funding for applications of GGE other than zygote editing.

As a result, we have to discuss GGE in a different manner. Instead of weighing the potential risks and benefits of a life-saving therapy, we should regard GGE as an elaborated infertility treatment. It is a procedure that allows reproduction partners to create a child to which both are genetically related.

This aspect is not in itself a valid argument against the use of GGE. As we have already discussed, the argument of reproductive autonomy might justify the use of GGE, apart from reasons of disease prevention. Reproductive autonomy is usually considered to be a strong argument. According to some authors, the principles of a liberal society demand, that whenever autonomy is restricted, there have to be substantial reasons (Harris 2005). Following this approach, the burden of proof is always on those who aim at restricting autonomy. Reproductive autonomy is mainly considered as a negative right that guarantees protection against any coercion to reproduce, or to select a certain reproduction partner. This view is a result of the experiences of political interventions into the area of reproduction in the 20th century. Especially in the first half of the century, several countries around the world implemented political programs to control reproduction, mostly through compulsory sterilization or marriage restrictions. The most notable example is the eugenics and euthanasia programs of the Nazi dictatorship in Germany. Apart from these historical reasons, reproductive autonomy is also a consequence of the civil rights and liberties that are crucial to liberal democratic societies.

Although there are strong arguments for reproductive autonomy as a negative right granting protection against state interference in reproduction decisions, this aspect seems insufficient when it comes to GGE. Apart from the question whether reproductive autonomy can also be interpreted as a positive right (which would imply the entitlement of reproduction partners to public funding), there is yet another point to consider. Reproductive autonomy usually implies a limited setting as well as a clearly defined number of individuals who are affected by a reproductive decision. Those affected are mainly the reproduction partners themselves and the child or children they want to create. When it comes to GGE, however, we are dealing with manipulations of the germline, which do not only affect the child created by using this method, but also possibly future generations. Therefore, the risk assessment must not only focus on the child, evaluating only whether the expected benefits of the treatment outweigh the risks on an ontogenetic level, but must also include the phylogenetic implications. In other words, we not only have to weigh the reproductive autonomy of the reproduction partners against the well-being of the child, but also against the interests of generations to come. This is an entirely different level of ethical evaluation. This does not necessarily mean an argument against GGE. It is merely to show, that as soon as we stop discussing the method under the false assumption of disease prevention, we find that the risk-benefit analysis changes dramatically. We can still debate whether we as a society are willing to take the severe intergenerational risks to fulfil the reproduction partners' wish for a healthy child to which both are genetically related. However, this debate is utterly different than discussing whether we should accept the risks of a life-saving therapy.

6 Conclusion

A sound ethical evaluation of GGE demands the appropriate conceptual framing of the procedure. A categorical argumentation, either affirming or opposing GGE without regard of the specific clinical context, is insufficient. The proper ethical evaluation of GGE demands a weighing of risks and benefits for each specific application. As we have seen, GGE offers various options for research as well as clinical applications. The focus of my analysis was on the latter, especially on the status of GGE as a means of disease prevention. This presumed therapeutic benefit is among the main arguments in favor of GGE. However, as a closer examination revealed, the assumption that GGE is a means of disease prevention is only true for one possible application, namely zygote editing. For the most part, GGE doesn't cure an embryo with a genetic condition, rather, it creates an embryo without that condition, which would not otherwise have been created. This makes most applications of GGE advanced techniques of assisted reproduction, which provide infertile reproduction partners, or those who suffer from a transmittable genetic condition, the opportunity to have healthy offspring that is genetically related to both of them. Therefore, these applications of GGE have to be discussed under a different paradigm. So far, GGE has mainly been discussed based on the question whether we are willing to accept certain risks of a procedure that provides great curative benefits. However, instead of focussing only on disease prevention, we should also discuss GGE as a means of wish-fulfilling medicine. In the light of this paradigm, we have to weigh reproductive autonomy not only against the interests of the future child, but also against societal interests. Since most applications of GGE are a matter of wish-fulfilling medicine, the question of public funding will have to be assessed differently. Even more importantly, we will have to reconsider the risk assessment. Therefore, we need a renewed public debate that differentiates between the only true preventive use of GGE which is zygote editing and the other uses of the method as a means of wish-fulfilling medicine.

References

Bayertz, Kurt. 1991. Drei Typen ethischer Argumentation. In *Genomanalyse und Gentherapie. Ethische Herausforderungen in der Humanmedizin*, Hans-Martin Sass (ed.), 291–316. Berlin/Heidelberg: Springer.

Baylis, Françoise. 2017. Human Nuclear Genome Transfer (So-Called Mitochondrial Replacement): Clearing the Underbrush. *Bioethics* 31: 7–19.

Callaway, Ewen. 2016. UK Scientists Gain Licence to Edit Genes in Human Embryos. *Nature*. doi:10.1038/nature.2016.19270.

Fogarty, Norah M. E., et al. 2017. Genome Editing Reveals a Role for OCT4 in Human Embryogenesis. *Nature* 550: 67–73.

Gómez-Tatay, Lucía, et al. 2017. Mitochondrial Modification Techniques and Ethical Issues. *Journal of Clinical Medicine*. doi:10.3390/jcm6030025www.mdpi.

Gyngell, Christopher, et al. 2017. The Ethics of Germline Gene Editing. *Journal of Applied Philosophy* 34: 498–513.

Harris, John. 2005. Reproductive Liberty, Disease and Disability. *Reproductive Biomedicine Online* 10: 13–16.

Harris, John. 2016. Germline Modification and the Burden of Human Existence. *Cambridge Quarterly of Healthcare Ethics* 25: 6–18.

Ishii, Tetsuya. 2015. Germ Line Genome Editing in Clinics: The Approaches, Objectives and Global Society. *Briefings in Functional Genomics* 16: 46–56.

Ishii, Tetsuya. 2017. Reproductive Medicine Involving Genome Editing: Clinical Uncertainties and Embryological Needs. *Reproductive Biomedicine Online* 34: 27–31.

Kang, Xiangjin, et al. 2016. Introducing Precise Genetic Modifications into Human 3PN-Embryos by CRISPR/Cas-Mediated Genome Editing. *Journal of Assisted Reproduction and Genetics* 33: 581–588.

Lanphier, Edward, et al. 2015. Don't Edit the Human Germ Line. *Nature* 519: 410–411.

Liang, Puping, et al. 2015. CRISPR/Cas9-Mediated Gene Editing in Human Tripronuclear Zygotes. *Protein & Cell* 6: 363–372.

Long, Chengzu, et al. 2014. Prevention of Muscular Dystrophy in Mice by CRISPR/Cas9-Mediated Editing of Germline DNA. *Science* 345: 1184–1188.

Ma, Hong, et al. 2017. Correction of a Pathogenic Gene Mutation in Human Embryos. *Nature* 548: 413–419.

Powell, Russell. 2015. In Genes We Trust: Germline Engineering, Eugenics, and the Future of the Human Genome. *Journal of Medicine and Philosophy* 40: 669–695.

Rubeis, Giovanni, and Florian Steger. 2016. Genome Editing in der Pränatalmedizin. Eine medizinethische Analyse. In *Jahrbuch für Recht und Ethik – Annual Review of Law and Ethics*, Bd. 24, B. Sharon Byrd et al. (eds.), 143–159. Berlin: Duncker & Humblot.

The National Academies of Sciences, Engineering, Medicine. 2017. *Human Genome Editing: Science, Ethics, and Governance*. Washington, DC: The National Academies Press.

Zhang, John, et al. 2016. First Live Birth Using Human Oocytes Reconstituted by Spindle Nuclear Transfer for Mitochondrial DNA Mutation Causing Leigh Syndrome. *Fertility and Sterility* 106: e375–e376.

III

Towards a Governance Frame in Dealing with Genome Editing

Who? What? How? Why? If You Don't Ask You'll Never Know …

On Criticism of the New Uproar about Germline Editing – Discourse Analytical and Socioethical Metaperspectives

Peter Dabrock

Keywords

Ethics, Genome Editing, Institutions, Reports, Responsible Research and Innovation, Trust

Abstract

Debates about moral, legal and political attributions of responsibility do not, as can be seen in the past, occur in a vacuum. Against this background, the following chapter does not directly address the pending ethical questions about human germline editing, but calls attention to several stages of social and ethical discourse. The goal, not simply in the interest of history, is to uncover the genealogy of the climate of the debate. After all, the hidden genealogy of discourses often contributes to the societal framing of a debate. Making these taken-for-granted assumptions explicit is of intrinsic value for further ethical debate. For possible argumentative short circuits, or abstractions, can be avoided through this making-explicit, and things forgotten or non-self-evident can be recalled in the advanced course of the debate. The focal point of this critical genealogy is the transition from the seemingly rather reserved statement of the so called Washington Summit of December 2015 to the about-face undertaken by the leading American academies (who were themselves involved in the Washington Summit) in the report they published in February 2017. Finally, how trust in science can be maintained beyond such perceptible interest-led transformations is explained in the concluding chapter.

M. Braun et al. (Eds.), *Between Moral Hazard and Legal Uncertainty*,
Technikzukünfte, Wissenschaft und Gesellschaft / Futures of Technology, Science and Society, https://doi.org/10.1007/978-3-658-22660-2_11

1 On the Ethical Relevance of Genealogical Reconstruction

Seldom has an article, published in what is surely a second-rate periodical, precipitated such a flood of media reports, commentaries and statements as the one submitted on March 30, 2015 (and accepted a day later) by the research group around Junjiu Huang of the Sun Yat-sen University in Guangzouh, entitled "CRISPR/Cas9-mediated Gene Editing in Human Tripronuclear Zygotes" (Liang et al. 2015). Using the blood disorder Thalassemia as an example, the researchers' ostensible aim was to demonstrate that (and how) a gene therapy performed on an early embryo could be carried out using CRISPR-Cas9. From the outset, the research group performed its experiments on embryos that had already been biotechnologically manipulated to the point of unviability. The success rate of their attempts was just as low as the attempts (also conducted in Guangzhou and published in April 2015) to alter the genetic make-up of humans for the purpose of producing a resistance to HIV (Kang et al. 2016). The results of both research groups were characterized by a high rate of side effects and inaccuracies. For this reason, many observers got the impression that the realization of germline intervention *in vivo* – in other words, the birth of genetically manipulated babies – is not to be expected in the foreseeable future. This was, for example, a widely shared assessment at the annual conference of the German Ethics Council (2017a).

Only scarcely two years later the German Ethics Council, along with many others, felt compelled to correct its assessment. In August 2017, under the direction of Shoukhrat Mitalipov of the Oregon Health & Science University in Portland/USA, an international research team was able to considerably reduce the side effects that occurred at such a high rate with the Chinese research group, and significantly increase the rate of effectiveness in its attempt at a germline therapy of an inherited heart disorder (Hypertrophic Cardiomyopathy), which is indicated at the earliest stage of human life (Ma et al. 2017). Even if in the meantime considerable doubt has been expressed about the meaningfulness of the experiments (cf. Egli et al. 2017), it can hardly be doubted that the possibility of bringing genetically manipulated babies to birth has moved into the foreseeable future.

This assessment is backed by the previous research. After all, the researchers from China and the United States frankly admitted that their aim is not only to provide therapy for a serious hereditary disease in an individual. Rather, the goal is to permanently prevent potential descendants of the respective embryo from carrying that disease. At issue, therefore, is a deliberate germline alteration and not only the accepted risk of an accidental one (as in radiological examinations or chemotherapy). The latter always exist in nature, mutations being the obvious and ordinary example. Yet, in contrast to mutations that happen by chance in nature, those which come about, and are then hereditary, through the human wish to deliberately steer their own evolution are ethically different. That is, they are to be classified as ethically relevant. For then, it is no longer possible to appeal to a pre-given fact of evolution. Instead, such an intervention constitutes an act which is not that of an individual, but happens in modern research projects, affects countless actors, is embedded in institutional settings, is linked to certain scientific, economic and political goals and is regulated by a complex arrangement of national and international hard and soft laws. For all involved individual and collective agents there are specific attributions of responsibility. Responsibility has to be taken for the realization of the chances hoped for as well as for the associated risks. That person also bears responsibility, who hinders research that, while not being without risk is nevertheless assessed as justifiable, the implementation of which, for example, could prevent serious illnesses and their associated harm to individuals and to the society.

Because such debates about moral, legal and political attributions of responsibility do not, as can be seen in the past, occur in a vacuum, the following contribution does not directly address the pending ethical questions, but calls attention to several stages of social and ethical discourse. The goal of this reconstruction, not simply in the interest of history, is to uncover the genealogy of the climate of the debate. After all, the hidden genealogy of discourses often contributes to the societal framing of a debate. Making these taken-for-granted assumptions explicitly is of intrinsic value for further ethical debate. For possible argumentative short circuits, or abstractions, can be avoided through this making-explicit, and things forgotten or non-self-evident can be recalled in the advanced course of the debate. Let us, therefore, look again at the beginning of the debate at the time of the publishing of the Chinese papers in 2015, and then in rough outline trace the way to the end of 2017. In doing so, the focal point of this critical genealogy is the transition from the seemingly rather reserved statement of the so called Washington Summit of December 2015 to the about-face undertaken by the leading American academies (who were themselves involved in the Washington Summit) in the report they published in February 2017. Finally, how trust in science can be maintained beyond such perceptible interest-led transformations is explained in the concluding chapter.

2 Some Pertinent Questions about the First Intensive Debate about Genome Editing in the Year 2015

While the leading science magazines *Nature* and *Science* knew but had rejected the article, possibly because of scholarly deficiencies or ethical considerations, pending its publication, the contents of the Chinese researchers' paper were known in the scientific community. Accordingly, both worldwide leading science magazines were able to prepare themselves for its publication and quickly offer skeptics a forum, advancing in this way to the head of the debate. Scientific organizations – among them the Berlin-Brandenburg Academy of Sciences and Humanities (2015) in Germany and a powerful consortium from Leopoldina, Acatech, the Alliance of Academies in cooperation with the DFG (Academy of Sciences Leopoldina et al. 2015), and in England a coalition ranging from the Academy of Medical Sciences to the Welcome Trust (The Academy of Medical Sciences et al. 2015) – practically raced each other in the issuing of statements. Even *The Economist* (2015) had already dedicated a title story to the topic ("The Age of the Red Pen") on 22.08.2015. There is hardly any greater media attention than this for a bioscientific topic. But, what was it that so excited, and still generally excites, everyone since mid-2015? There had been an awareness of the significance of the bacterial defense mechanism CRISPR-Cas9 – as a relatively easy to apply, precise, effective and cheap genome editing tool for the biosciences – going back to the 2013 publication of the groundbreaking article by Jennifer Doudna, Emmanelle Charpentier and their team (Jinek et al. 2012). The media excitement is thus obviously explained by the highly symbolic and ethically relevant application of genome editing in the area of possible germline interventions, which the Huang Group showed for the first time to not be completely impossible.

The following characteristics are detectable in the first debate from mid-2015: Hitherto regarded by many as taboo, most statements initially use the occasion of the application of the gene scissors in germline interventions to acknowledge their scientific significance. Thus, the hymns of praise virtually fall over each other with their superlatives: "powerful" (The Academy of Medical Sciences et al. 2015), "revolutionary" (Berlin-Brandenburg Academy of Sciences and Humanities 2015), "tremendous value" (The Hinxton Group 2015). Moreover, there is not one statement which, in the face of the newly emerging possibility of a germline intervention with the help of the CRISPR-Cas9 procedure, would not call for an intensive public debate. In one and the same statement, namely that of The Hinxton Group, the role of science can be characterized both in terms of the deficit model[1] (officially

1 The deficit model plies the schema of an asymmetrical polarity: Here an informed science – there the (supposedly) "irrational public excitement".

believed to have been long since overcome) and in terms of a participative system that stands for public engagement with science, open for civil scientific considerations. This is an ambivalent gesture that informs not a few scientific organizations in their dealings with the public, which is often eyed with suspicion. Provided as grounds for demanding public debate was either (surely primarily motivated by self-interest) the threat of the loss of public trust in science ("public trust in science ultimately begins with and requires ongoing transparency and open discussion" [Baltimore et al. 2015, 38]), or the result (instantly interpreted in dystopian terms) that via genome editing the possibility of genetic enhancement and the irreversible transformation of the genetic basis of humankind has realistically appeared on the horizon (ETC Group 2016).

That the scenario of a possible germline intervention with the goal of clinical application should give occasion to worry, and results in public debate, is hardly surprising. Astonishing, rather, is who is taking a prominent stance as the bearers of moral indignation: a succession of researchers not otherwise known as wallflowers of biotechnological progress – in other words, researchers who regularly participate in stem cell research that makes use of embryos.[2] If, following Niklas Luhmann, ethics is a "warning against morality" (Luhmann 1990, 41), or put less provocatively, if ethics has to establish a criteria-led distance to moral communication, then, with a view to the first peak of the genome editing debate in mid-2015, it may rightly pose the following questions: Who? What? How? Why?[3]

2.1 Who?

The protagonists of the international debate, e.g., George Church (2017), Francis Collins (2015), or David Baltimore (et al. 2015), were not hitherto noted for wanting to hinder research, nor had they distinguished themselves by a marked engagement in the protection of life. When those for whom research otherwise never moves fast enough bring out the big ethical guns here, warning about incalculable dangers, should not one ask which interests or causes are guiding them? (On this rhetorical question, cf. Braun et al. 2016) Did some of those who played to the gallery as guardians of morality perhaps, in truth, feel peeved that this scientific breakthrough, so intensely followed in the media, was proclaimed in China and not in the USA or the

2 Cf. only the authors lists of both following contributions: Lanphier et al. 2015; Baltimore et al. 2015.

3 In the German context, this four Questions function as the essential part of the Sesame-Street theme song.

UK?[4] For the marketing strategy of scientific innovations, in America significantly stronger than in Europe, such a media-hyped biotechnological success in a another country may have been a considerable outrage, being perhaps also associated with substantial financial disadvantages. Such questions were asked with justification in 2015, because research on in vitro embryos was not and is not per se forbidden in the USA or in the UK. Indeed, it has been explicitly allowed in England since 2001, and is at least possible in the USA (although not publicly supported at the federal level or in many states). Considering that the Chinese had not used viable embryos, there is no detectably exceptional or scandalous particularity, where the "use" of embryos is concerned, in the Chinese experiments vis-a-vis the prevailing legal framework in the UK and the US. Accordingly, one must be permitted to ask critical questioning of this coalition of sceptics.

If one defends the apprehensions of these sceptics, by pointing out that it was not only the embryo research which produced moral uproar, but the intentional intervention into the human germline, then it should have been pointed out at the time, that this alleged breach of taboo had already been carried out in England. After a longstanding, intense and emotional debate, mitochondrial transfer was allowed in England in 2015. To argue that mitochondrial transfer is not a matter of intervention in the germline because there is only a quantitatively meager part of the genome (and besides, not the nuclear genome,) affected here would be an argumentative fallacy. Neither arguments of quantity nor of locality are valid for the dismissal of this classification. Even if the consequences of a genetic-surgical intervention in the nuclear genome result in a considerably more specific intervention depth, in the emphatic words of Johannes Rau (which, however rightly criticized, intimate the paradigm change): With mitochondrial transfer, the "Rubicon" (Rau 2001) has already been crossed. It seemed all the more self-unmasking – viewed from the outside – when the sharp critics around David Baltimore explicitly directed their concerns only toward nuclear genome surgery, while completely screening off the scientifically obvious reference to mitochondrial transfer (Baltimore et al. 2015). In any case, such a screening-off is not justified by the familiar motto: "Only complexity can reduce complexity" (Luhmann 1995, 26). For in a systems biology approach, which is anything but foreign to the aforementioned authors, this connection between changing the nuclear genome and mitochondrial transfer is more than obvious, and therefore, also requires ethical reflection. Or, formulated the other

4 The hermeneutic of suspicion, that the Americans could be "jealous" of the scientific break-through, would most likely not be attributable to the quality of the scientific publication (which was lousy), nor to the reputation of the researchers (who were almost unknown), but to the worldwide media attention that the Chinese managed to grab.

way around, undertaking the strict decoupling of both genetic technologies needs explicit ethical justification. Where this does not happen, such an unfounded, stark assertion of the possibility of their decoupling is in need of criticism. In any case, by its call for contextualization, this criticism of the critics of germline intervention issues in the warning to stop styling the Chinese experiments in terms of a cultural ditch between conscientious Western and unconscientious Eastern hemispheres.

Conversely, however, it also holds that only naive romantics are surprised that protagonists have their own interests. Interests may make positions suspect, but they do not deny good arguments, which, if they really are good, retain their validity even after the interests have been uncovered. The already mentioned call for public debates only appears to be sensible if – beyond the (in my opinion, insufficient) formal communicative signal to give the public space and time to come to terms with, and accept, new challenges – the respective supporters of a new biotechnology also introduce material criteria into the debate, explaining why which measures are to be taken, and why the factual and social benefits not only outweigh the disadvantages, but also do not exceed a certain degree of risk (completely independent of a possibly positive cost-benefit analysis). At the same time, until the beginning of 2017, most statements are for the prevention of (intentional) germline interventions as the self-evident goal of such public discussion. In mid-2015, however, the statements of The Hinxton Group and the Academy of Medical Sciences already depart from this broad (for many seemingly self-evident) consensus. They judge the reservation about germline intervention solely by the currently unclear risk assessment. In the international sector, however, the strict rejection of germline intervention is formulated more emphatically, e.g., in the Oviedo Convention's explanatory report on art. 13, which deals with interventions into the germline (Council of Europe 1997). As reasons for strict reservation it expresses concern about putting the foundations of the human species at risk. While, where international law is concerned, the Oviedo Convention is the only bioethically relevant binding contract, the binding character of the UNESCO Declaration, which also rejects germline interventions and declares the genome to be the symbolic heritage of humanity (UNESCO 1997), is much lower. Nevertheless, it too underscores the prevailing, broad consensus against germline interventions.

2.2 What? How?

What measures do the individual sceptics suggest taking in the face of the new possibilities of intervention into the human germline? Time and again the call is actually made for a moratorium. Beyond the simple mention of this watchword, the statements from 2015 differ considerably in the range and goal of the determination of such a suspending interruption. Besides an individual voice in the *Washington Post,* of the widely respected statements only the group around Lanphier (et al. 2015) called for a provisional stoppage of the whole of genome surgery (probably more precisely: in the human domain): "[A] voluntary moratorium on all [!] kinds of genome editing in the scientific community could be an effective way to discourage human germline modification and raise public awareness of the difference between genome editing in somatic and in germ cells" (Lanphier et al. 2015, 410–411).

That nothing else is adduced for the justification of such a drastic step than internalizing a scientifically important distinction is, however, surprising. And, if disputed, it would hardly stand up in court. The previously mentioned individual voice in the *Washington Post*, the policy advisor Vivek Wadhwa, sees the scientific world, in light of our entrance into the technological age of "backyard synthetic biology", virtually coming apart at the seams with the arrival of CRISPR-Cas9: "No one is prepared for an era when editing DNA is as easy as editing a Microsoft Word document. The government does not have any regulations on editing human DNA. The ethical concerns have not been fleshed out. There is no centralized risk-management inventory, listing which labs are doing what with CRISPR. It's all rather terrifying. Rarely do I argue that a moratorium on technological progress is the prudent course. But the stakes in the case of CRISPR are so high that I believe a blanket moratorium is the only course" (Wadhwa 2015).

Most other statements, by contrast, reduce the object of the moratorium solely to the immediate use of gene surgery for intervention into the human germline in view of possible reproductive purposes. Put differently, it is primarily security and risk assessments which, if at all, could justify a moratorium limited to *this* application. There are at least some who bring themselves to say that the time of the moratorium, limited to germline application, should be used to publicly address fundamental questions about the sense and purpose of such researches.

2.3 Why?

Most of the time stimulation of public discourse is already viewed as a sufficiently significant ethical goal. Refreshingly, the statement of the Berlin-Brandenburg Academy of Sciences and Humanities (BBAW 2015) differs from most other statements, which, in an astonishing and especially telling way, only produce empty spaces at this point. In the BBAW statement, by contrast, arguments are presented for and against germline intervention. For example, a case can be made for the possibility of the "elimination of a severe risk of illness for potential descendants". In view of this considerable good the theoretically posable counterargument of the artificiality of the procedure does not at all have the desired effect. With reference to individuals, the "violation of the right to bodily self determination and integrity" could be invoked against germline intervention. In the following adversative formulation, "although the individual does not even exist yet", the BBAW statement indicates straightaway their skeptical attitude toward this argument. The same goes for slippery slope arguments, which, on the whole, are always classed as weak. Yet these need to be carefully distinguished (which does not happen in the BBAW statement) from so called ex-ante-scenarios, which, unlike slippery slope figures, do not simply play up unrealistic horror scenarios. In contrast to the bursting-dam-rhetoric of slippery slope arguments, ex-ante-scenarios gain argumentative force, if (or because) they develop scenarios with clearly verifiable, and from the immediate present, prolongable conditions in the foreseeable future. This could have, indeed, should have been differentiated in the BBAW statement. The impression arises, inasmuch as this does not happen, that every attempt at the careful development of an ex-ante scenario already succumbs to the ethical verdict that rightly applies to slippery slopes. But that is not the case!

According to the tenor of the BBAW statement, the further difference criteria for the ethical and critical assessment of genome surgery (the therapy/enhancement distinction; protection of the human species; technological self-design as a part of the historical heritage of human persons) are unlikely to be sustainable for a categorical rejection in the bioethical discourse. That is to say, if for example one criticizes enhancement on the grounds that it would thereby embrace a perfectionistic lifestyle, then one would regularly be exposed to the objection: "All that does not justify prohibiting (the implementation of) the opposite opinion!" According to this stance, it is the prohibition and not the permission which bears the burden of proof. This is purportedly the case not only constitutionally, but also for the principles of political philosophy or social ethics, which the plural society must take seriously.

Even if, in the short ethical considerations of the BBAW, one misses substantial discussions about crucial ethical criteria and interpretive patterns, and their application to genome surgery,[5] in light of the liberal approach that often predominates in scientific and research ethics, the fundamental skepticism remains about whether

5 For example: Precaution, reversibility, retinity, the problematic constriction of the open future (Feinberg 1980), the problematic cultural signal of a biotechnological project development of human existence, the threat to traditional cultural imaginations like solidarity with vulnerable (groups of) persons, the loss of the basic preservation of the conditions of a human way of living, an enlightened use of the precautionary principle, the problem of complicity, the responsibility for acts of commission and omission.

such criteria can, on the whole, justify a prohibition of germline interventions that is not only based on considerations of security and risk. For all these considerations, so the expectation, much time should remain. However, following an intermittent *ritardando*, the ethical (and scientific and scientific policy) situation has in fact – it can be put so emphatically – changed dramatically over against mid-2015.

3 Illusory Reassurance through the Washington Summit at the End of 2015

The "International Summit on Human Gene Editing", the so called "Washington Summit" of December 2015, which was staged with great media effort and corresponding media presence, understood itself to be the scientific community's reaction to the Chinese experiments. Because the organization of the summit was headed up by the national science academies of the countries that had distinguished themselves by particularly intensive use of embryos in research, of whom further intensive engagement was expected using genome editing (namely the USA, Great Britain and China), great hopes were placed in this high-level conference. Yet the expectations of some, that a moratorium would be explicitly proclaimed here, were not only too high, but surely also naive. How, in the face of the scientific, economic and biopolitical dynamic, could one seriously expect that? Nevertheless, the concluding communiqué exhibits a keenly reserved assessment of the possibility of germline intervention with the aim of clinical application, that is, with the intention of bringing about the birth of a genetically manipulated human. For it states: "It would be irresponsible to proceed with any clinical use of germline editing unless and until (i) the relevant safety and efficacy issues have been resolved, based on appropriate understanding and balancing of risks, potential benefits, and alternatives, and (ii) there is broad societal consensus about the appropriateness of the proposed application. Moreover, any clinical use should proceed only under appropriate regulatory oversight" (The National Academies 2015).

Owing to the further development in the global discussion about germline interventions, two points are to be considered as non-self-evident: 1. Genome editing in clinical use is regarded as fundamentally "irresponsible". This characterization will only be abolished when both scientific and social requirements are completely met. That not only scientific questions need to be resolved, but also that compliance with the societal dimension is considered a necessary condition for overcoming the fundamental irresponsibility of germline interventions by means of genome editing, is unusual. 2. The Summit goes out on a limb, because it does not consider the social conditions already to have been met when an extensive societal debate has taken place, but only when "broad societal consensus about the appropriateness of the proposed application" has been found. Given the highly plural global society, in which the *signum* "clash of cultures" (Huntington 2003) is experiencing a renaissance, the fulfillment of such requirements is to be viewed as well-nigh utopian. It is hard to believe that the authors of the statement were unaware of this. If one were to take the summit at its word, then such a demand would *de facto* be tantamount to a moratorium, even if the word does not appear in the document. Thus, it is not surprising that the end statement soberly asserts: "At present, these criteria have not been met for any proposed clinical use: the safety issues have not yet been adequately explored; the cases of most compelling benefit are limited; and many nations have legislative or regulatory bans on germline modification. However, as scientific knowledge advances and societal views evolve, the clinical use of germline editing should be revisited on a regular basis" (The National Academies 2015). One can conclude from this, even though the keywords hoped for by some were missing ("moratorium", "ban", "pause"), that the concluding communiqué was nevertheless a strong message both with regard to the call for criteria and the exceedingly sober assessment of the scientific risks and regulatory ambiguities. And, in view of the participating organizations, one could have promised peace on the front of the displacement of hitherto established boundaries, while at the same time hoping for society-wide intensified debates in many countries. Although numerous events were organized (for example, the German Ethics Council made "Access to the Human Genetic Makeup: New possibilities and their ethical assessment" the theme of its annual public conference [German Ethics Council 2017a]), and in view of its possible significance for the self-understanding of humanity, the topic did not find any more outstanding response in 2016. Perhaps people allowed themselves to be calmed down too much by the result of the, admittedly only pragmatic, and yet clear and keen reservation of the Washington Summit with respect to a responsible employment of genome editing for the purpose of clinical application.

Yet already in February 2016, i. e., almost one and a half years before the Mitailipov Group's publication, a certain biopolitical turn could have shown that the peace was illusory. At that time, the British Human Fertilisation and Embryology Authority gave a team of researchers, around Kathy Niakan from the Francis Crick Institute in London, permission to conduct germline intervention experiments. In contrast to the Chinese experiments, and the researches published later in August 2017 by the Mitalipov Group, they did not directly aim at therapeutic applications, but had in mind basic research in the area of developmental biology. Niakan's attempts were probably not perceived as something special in England, since permission had of course already been granted for a small germline intervention through the authorization of mitochondrial transfer. In doing so, the fact that it was considered permissible to allow this germline intervention was deliberately overlooked, precisely because it was clearly not a large intervention. In any case, Niakan's experiments did not incite any considerable debate.

4 The Game Changer: The Report of the American Academies

The actual game changer, in the whole debate about germline intervention via genome editing, was the joint-report of the National Academy of Sciences and the National Academy of Medicine Committee, "Human Gene Editing: Scientific, Medical, and Ethical Considerations", published in February 2017. With only a small readjustment of language, a biopolitical paradigm change broke fresh ground (as can be seen in its consequences), preparing the way (ethically as well as in terms of scientific policy) for the publication of the Mitailipov researches (which were of course certainly already known in their working stages). While the Washington Summit had still spoken of a fundamental irresponsibility of clinical germline interventions by means of genome editing, which could only be abolished via the high hurdles of the scientific (the solution of safety issues and efficacy problems) and societal (a "broad consensus") requirements, the report of the academies reversed the burden of proof. Now it suddenly stated that germline interventions for a clinical purpose are "permissible" (The National Academies 2017, 135), if (at least) a long list of ten criteria are fulfilled. On the one hand, the Washington Summit's very general call for attention to substantial requirements of security and effectiveness is made more concrete with many of these conditions.[6] On the other hand, regulatory and governance demands, like those for transparency, privacy and oversight mechanisms, are made more concrete. Especially the demand for a "long-term, multigenerational follow-up that still respect[s] personal autonomy" (The National Academies 2017, 135) stands out as a remarkable feature. For it makes clear that

6 The following demands serve this goal: "restriction to preventing a serious disease or condition; restriction to editing genes that have been convincingly demonstrated to cause or to strongly predispose to that disease or condition; restriction to converting such genes to versions that are prevalent in the population and are known to be associated with ordinary health with little or no evidence of adverse effects; the availability of credible pre-clinical and/or clinical data on risks and potential health benefits of the procedures" (The National Academies 2017, 7–8).

the authors are trying to keep an eye on possible epigenetic effects, which may only appear after two to three generations. I expressly formulate it thus ("are trying to keep an eye on"), for if one were really to take such epigenetic aspects seriously, as was previously done in a joint position statement of the American Society for Gene and Cell Therapy (ASGCT) and the Japan Society of Gene Therapy (JSGT) (Friedmann 2015) (whereas George Church [2017] discounts them as belonging to normal life-risk), then one could hardly expect that only the biotechnological risks would be resolved in the foreseeable future.[7] Obviously, taking these hurdles into account (which had still induced the authors of the Washington Summit to exceeding reservation) appears to be only a slight thing for the authors of the Academies reports, since their overall pronouncement is "permissible". The word choice of the central recommendation 5-1 is carefully prepared in the text, in that firstly, in a dubious (because either trivial, or a for a scientific description impermissibly normative) formulation, the realistic possibility of germline interventions is indicated. Then, resting on this doubtful premise, both a prohibition approach with a suspension of permission, and an approach that initially assumes a fundamental irresponsibility (like the statement of the Washington Summit) are explicitly rejected, in order then to proceed to the aforementioned, recommendation-informing "permissible". Despite the still discernibly unclear risk assessment – which is primarily based on the (for a long time) yet incalculable epigenetic effects, and which gives good reasons (following the precautionary principle) for leaving the burden of proof with the supporters – a reversal of the burden of proof is nevertheless initiated. It focusses on a quasi defense-rights oriented, science-friendly and science-centered understanding of freedom, which, to be sure, has to take aspects of safety, security and efficacy into consideration, but which, in open cases of doubt, scarcely considers a precautionary approach (there is no mention at all of a precautionary principle).[8] Astonishingly, or tellingly, the word "precaution" only appears in the report in the appendix about international regulations (showing that it could, in principle, have been on the radar). It does not, however, appear in the main text, whether in its recommendations or even in its overarching principles, which should guide an ethically responsible use of genome editing.

---

7 On the ethical and regulatory problems (which have no foregoing example) of such long-term multigenerational follow-ups, cf. Cwik 2017.

8 On the distinction between the rather proactive, regulations-oriented European precautionary principle and the rather reactive, pragmatically oriented American precautionary approach, which threatens with liability law, and otherwise relies on the self-commitment of the agents, cf. Bourguignon 2015.

In addition to the shift of the distribution of the burden of proof, there is a second shift, which can hardly be considered less dramatic. If in the Washington Summit there was still talk of requiring a "broad consensus" as the *conditio sine qua non* for a non-irresponsible use of genome editing for clinical germline interventions, it has been completely lost here. Now a public debate is indeed thought necessary, and it is even admitted that there can be quasi-deontologically, and not only consequentialistically motivated approaches against germline intervention (although the report does not consider that there can also be strongly motivated consequentialist reservations, that, because the expected negative consequences can scarcely be temporally, factually and socially rectified, could take on a deontological character). Nevertheless, the contribution of the public is now limited to a "broad on-going participation and input by the public" in the "continued reassessment of both health and societal benefits and risks". In short, there is no longer talk either of deontological arguments, or of a genuine participation in decision making, which still shone through in the "broad consent" of the Washington consensus. Thus, a genuine "public engagement with science" about such a fundamental question is rejected. Even in such a weighty question as this, which may impact all of humanity, the text no longer gives the impression that it understands science as embedded in society.[9]

In the face of this direction change (light in its tone but drastic in its substance), via the reversal of the burden of proof and a reluctance to allow the public to have a say in questions of governance, other inconsistencies are almost completely overshadowed, although they are of enormous importance for the matter at hand: The point of greatest weakness may be the ambiguous and vague formulation – which shows how shady the argument is at such an important point – that a necessary condition of a "robust and effective regulatory framework" (and the first to be mentioned at that) would have to be the "absence of reasonable alternatives". But what does that mean? Since the premise is set up in the introduction, that "[i]n some situations heritable germline editing would provide the only or the most acceptable option for parents who desire to have genetically related children while minimizing the risk of serious disease or disability in a prospective child", one may assume that the framework for "reasonable alternatives" is the desire for a genetically related child (cf. Hyun et al. 2017). However, two alternatives are withheld, which could both very well be described as reasonable.

Put in the form of a question: Who actually says that there is no reasonable alternative to a "genetically related child"? Is adoption per se not a reasonable al-

9 Cf. the similar critique of one of the authors of the statement of the Washington Summit, Francoise Baylis (2017).

ternative? These rhetorical questions, which wish at least to broaden the horizon of the report, lead to the follow-up question debated for decades in bioethics: Is there a right to a genetically related child? Which right is meant? A right of claim or a right of defense? But if "merely" the (in liberalism easier to justify) right of defense is meant, the question still arises: Is it absolutely valid, or does it sometimes and in some ways need to be weighed against other rights? Yet, before this question can be answered, it needs to be clarified whether – under the (hypothetically accepted) assumption that only ways which enable a genetically related child are counted as "reasonable alternatives" – there are still other "reasonable alternatives". It has been pointed out time and again that PGD could be such an alternative (cf. Hyun et al. 2017). If it is set aside that PGD is itself highly controversial, but is an option even in Germany (subject to certain conditions, being in many respects treated and assessed like other PNDs), then it is not possible (at least in the US, where negative rights and freedoms are paraded around like a monstrance) to say that PGD is not a "reasonable alternative". In any case, this will have to be said for a majority of those possibly affected.

Yet the debate would actually need to go further. The plausibility (or implausibility) of whether PGD can be viewed as a serious, "reasonable alternative" depends decisively on the perspective taken. If one exclusively takes the perspective of the would-be parents, then the trickling away of time, which, through unsuccessful attempts, allows the window of opportunity to close on their own reproductive capacities, is to be considered "reasonable". For these couples perhaps, a situation develops which gets evaluated as "the absence of reasonable alternatives" to germline editing. The situation is different for a society that arrives at the assessment that only very few couples, if any, can be offered germline editing as a feasible, not to mention a "reasonable", alternative. From the perspective of the society, then, a balancing of interests can very well be undertaken with regard to whether societal interests, like distribution of research money, or endangerment of societal peace, do not outweigh this quantitatively minor wish (cf. Hyun et al. 2017). However, these considerations must not (from an ethical point of view) do harm to an individual's dignity. This, in turn, leads to the complex question of how the right to reproduction is to be interpreted. However this question is answered, it should have become clear that, already at the first hurdle for the permission of genome editing, the "absence of reasonable alternatives", there is a considerable need for discussion. The great crux of The National Academies' report is that it only assigns the questions paradigmatically mentioned here, at best, a "social risk/benefit assessment". It goes unrecognized that in germline editing fundamental questions, indeed it requires such emphatic wording: questions of humanity as a whole, are addressed, which cannot be dealt with in passing via a "continued reassessment of both health and societal benefits and risks, with broad on-going participation and input by the public" (The National Academies 2015, 8).

Still many further questions could be cited, extending far beyond the report's undue, purely scientific focus. Here are a few paradigmatic examples: Do we (whoever "we" refers to and represents) want to undertake such changes? Can a goal (here the therapy of serious diseases) remain morally good, if the ways to it (here an indefinite time of hazardous experiments on later persons, who would not be able to consent) are morally questionable to a considerable degree? Where do "we" draw the line for such germline changes? Thousands of monogenetic hereditary dispositions alone are known. Which gene variants should be "erased"? Do we want these sorts of manipulations, even if they provide a template for more far-reaching perfecting of the human? What risks do we want to burden later generations with, knowing, through insights gained in recent years in systems biology and epigenetics, that some consequences of genetic changes first appear in children's children, who also cannot be asked for their informed consent? Do such therapies foster an image of a perfect human, or a human to be genetically improved upon, by the same token lowering sympathy for vulnerabilities? Will natural conception not be discontinued, simply because "gene-corrections" will anyway only be affordable for a rich upper-class? And can even hardened rationalists pass over the fact that surely billions of people feel religious or world-view motivated unease at such profound interventions into what is given? For all these questions, which could easily be expanded,[10] there are controversial positions. At the level of global politics, they are currently being discussed in expert committees (for example, in October 2017 in the Council of Europe [2017] on the occasion of the 20th anniversary of the Oviedo Convention, or in UNESCO's International Bioethics Committee [UNESCO et al. 2015]). For this reason, a devastating signal is sent when, as with The National Academies (who were themselves involved in the Washington Summit), such a perceptibly interest-driven and partially incoherent paradigm change is undertaken, which excludes the populace from a question that pertains to all humanity.

It must be stressed that, contrary to The National Academies' report, no matter how one answers the questions, they must not only be debated on the scientific level, but also in the spheres of civil society and politics – and at all conceivable levels simultaneously. The phrase "the genome as the heritage of humanity" (UNESCO 1997), weak at the level of international law, and yet strong in substance, captures the drama of the task in light of the most recent research. Thus, it was with justification that, in its ad-hoc recommendation on germline intervention,

10 Cf. for a more comprehensive list of questions, German Ethics Council 2017b.

the German Ethics Council (2017b) called for this broad participation of society and a general policy in the questions about germline editing. It did so unanimously, although the committee reproduces within itself the plurality of society, and although one could not conclude, from the publications of individual council members, that an equally unanimous rejection of germline editing might lurk behind this unanimous appeal.

5 Trust in Institutions is Crucial, but Extremely Endangered

If one is ethically aware that each of the arguments and criteria can be individually refuted, where the application of genome surgery for the purpose of clinical applications is concerned, then one can at least pose the metaethical question, whether the interaction of these arguments, each individually remaining in a hypothetical state, is sufficiently weighty as a whole, that not only arguments of technology assessment are permitted in the assessment of germline intervention ("the whole is more than the sum of its parts"). The democratic constitutional state (which is not only a liberal, but a democratic constitutional state) is of course also otherwise familiar with value judgments, allowing (sometimes gladly, sometimes grudgingly) the sovereign, the people, to legally define, or at least canalize, moral attitudes. Beyond political liberalism (which, of course, is always upheld in certain moments by those who find it useful just then for their maximum preservation of freedom), this ethical contamination of law, which is of course an expression of democratic decision-making, will hardly be surprising.

It is one of the insights of ethics' perception and interpretation of reality, that the debates about biotechnological breakthroughs seldom find their fulfillment in juridical risk analysis, or even in strictly ethical deliberations. Time and again, scientists as well as those interested in questions of governance are surprised and appalled that the "untamed" public (or at least parts thereof that are effective in the media) does not want to bow to, much less break out in shouts of hurray for the supposed better scientific argument in favor of the ever so sensible biotechnological development, and that even the most patient argumentation falls flat in the end. One only needs to recall the debates about the so called "green genetic engineering", experimental research on animals, or atomic energy.

Why is this? Can and should one do nothing at all, or, as a scientist or scientific organization, throw in the towel in total resignation? Taking various cases as examples (and genome surgery could be in danger of the fate of the aforementioned technologies), one can make an impressive study of blockades set up by the public, or individual publics, against certain technological developments, up to the point where those developments founder. In my opinion, Klaus Kornwachs has plausibly indicated a first level of this foundering: "Acceptance research, as a special field of sociology and in particular technical sociology, has discovered that the critique of technical systems, technical developments, or technical facilities sites, does not aim so much at their technical substance as such. The enemy is not the technology itself (even though there is literature that is fundamentally critical of technology, which some spokespersons have invoked), but the societal, economic and cultural changes, which they fear could inevitably accompany the respective technology." (Kornwachs 2013, 64) To cite Günter Anders, such (well-nigh Heideggerian) *angst* (and not simply *worry*) articulates namely, "that we are no match for the perfection of our product; that we produce more than we can represent or answer for; and, that we believe we are permitted, or rather obligated, nay compelled, to do what we are capable of doing." (Anders 1980, VII) It needs pointing out, precisely in the age of so called fake news[11] that (viewed sociologically) it is a secondary matter whether this unease corresponds to a reality describable by scientific criteria. Especially when this unease takes on speech forms with connotations to mythological ("Pandora's box"), religio-cultural ("playing God") or literary ("Frankenstein") motives, the observable mood approaches what Luhmann described as "protest communication" (Luhmann 1996) – i.e., a fundamentalist attitude averse to technology (that this mood then avails itself of all manner of technical media should only be mentioned as an especially interesting paradox – an incongruity, however, that does not bother the respective agents). This protest communication articulates a cultural unease (cf. Ried et al. 2011) that interacts with the projected transformation dynamics, and while it is not absorbed by them, it can nevertheless decisively co-determine their acceptance or non-acceptance. Kornwachs (2013, 108) construes this pattern of communication by arguing that the communication of the prognosis influences that about which the prognosis is. He thereby takes up, for the question at hand, a pattern referred to in social psychology as the Thomas theorem: "If men define situations as real, they are real in their consequences." (Thomas et al. 1928, 571–572) Beyond the (ultimately global) social assessment of germline intervention by means of genome editing, which needs to be made on different levels, the further basic question arises: How can science deal with such developments, which it perhaps deems mere irrational panic, in a prospective and

11 It is important to note that such fake news have always existed, but the spread and effect of which, via the effect of the digital media society, described in terms of echo-chambers and filter bubbles, is more profound than in previous decades.

constructive manner? That science (not necessarily *de jure*, but de facto) has an obligation to be aware of and deal with societal expectations is shown by recent surveys (Pew Research Center 2015; Wissenschaft im Dialog 2017). They make clear that trust in science (presumably conditioned in part by the general loss of trust in expert systems, of which, however, science is presumably one of the most important), although still at a high level, has nevertheless dropped significantly in recent years (Bertelsmann-Stiftung 2017). If ethics wants to comment on this, then it is necessary to include also modes of social ethics as well as governance theory (cf. Braun et al. 2017). Even if in the process the already outlined security, risk and technology assessment insights, as well as the juridical and ethical criteria, are not only not to be relinquished but rather to be taken account of, the acceptance of new scientific breakthroughs still essentially depends on whether and how the participating scientists, consortia, or the organizations representing them, have engaged in continual trust-building and mistake management. Studies, including our own, have shown that many people can scarcely evaluate the contents of scientific research. However, they certainly have a feeling for the high complexity of individual research projects as well as the complicated interdependence and associated sluggishness of certain scientific developments. As we were able to demonstrate in our study of emerging biotechnologies, they confer a kind of "granted loan" – that is, if you will, a trust fundamentally based on continuity, yet not without precarity (cf. Starkbaum et al. 2015). At least here in Europe flashy announcements of scientific breakthroughs, with the usual promises of salvation, are not only less and less believed, they are categorized as unnecessary, and even counterproductive for the reputation of science. In Germany and Austria more than in other European countries, but even in the USA, this tendency can be observed with reference to synthetic biology. Once massively shaken, basic trust in research will not be able to refute the impression that there is a lack of transparency, of obvious entanglement of predominant interest, or even the suspicion or accusation of fraud and abuse. As in a dialectical counterblow, this intensive trust can then turn into sustained mistrust that can have the effect of endangering progress. Such fears have also been voiced about genome editing (cf. Dance 2017). It is true, empirical research cannot at the moment confirm this trend for the assessment of genome editing (Gaskell et al. 2017). However, in the sense of a social concept of precaution, which also seeks to anticipate and responsibly shape the social consequences of a technology (cf. Wolpe et al. 2017), one should not be content to rest on this empirical result. Because (at least in much of Europe) public research institutes primarily stand for positive values, they are able to regularly book the highest advance of trust (according to the continuing tendency of the Euro-barometer in recent years [cf. European Commission 2016; European Commission 2017]). Only with this continuously cultivated, trust-generating framework (into which the combinations of technology assessment, juridical and ethical argumentation strategies, often considered by science to be sufficient, will be introduced as necessary, but precisely not as sufficient) will it be possible for science to meet the – at first glance surprising or annoying – waves of skepticism with a fair amount of success. Dealing with these apparently irrational waves of skepticism is painstaking work (as participation is always painstaking in general). However, it is an expression of the fact that the law and the constitution are only cultivated and maintained in a democratically constituted polity, which must not only bear with, but also promote the advantages and disadvantages of its own constitutive deliberation. In societies shaped by a democratically constituted culture of civility, it is highly probable that there will be a corridor for responsible research and innovation accompanied by a wide spread societal consensus. It is precisely in this question – which concerns humankind as such – whether genome editing for the purpose of germline interventions can be used, that it will be shown to what extent science is embedded in society, and to what extent it can gain trust and be deserving of trust.

References

Academy of Sciences Leopoldina, et al. 2015. *The Opportunities and Limits of Genome Editing.* Halle: Academy of Sciences Leopoldina et al.

Anders, Günther. 1980. *Die Antiquiertheit des Menschen. Über die Seele im Zeitalter der zweiten industriellen Revolution.* München: C.H. Beck.

Baltimore, David, et al. 2015. A Prudent Path Forward for Genomic Engineering and Germline Gene Modification. *Science* 348: 36–38.

Baylis, Françoise. 2017. Gene Editing: Where Should We Draw the Line? https://impactethics.ca/2017/10/25/gene-editing-where-should-we-draw-the-line/. Accessed 29 March 2018.

Berlin-Brandenburg Academy of Sciences and Humanities (BBAW). 2015. *Human Genome Surgery – towards a Responsible Evaluation of a New Technology: Analysis by the Interdisciplinary Research Group Gene Technology Report.* Berlin: Berlin-Brandenburg Academy of Sciences and Humanities.

Bertelsmann Stiftung. 2017. Social Cohesion in Germany 2017. English Summary. https://www.bertelsmann-stiftung.de/fileadmin/files/user_upload/LW_Abstract_Social_Cohesion_in_Germany_2017.pdf. Accessed 29 March 2018.

Braun, Matthias, and Peter Dabrock. 2017. Mind the Gaps! Towards an Ethical Framework for Genome Editing. *EMBO Reports.* doi:10.15252/embr.201745542.

Braun, Matthias, and Peter Dabrock. 2016. "I Bet You Won't": The Science–Society Wager on Gene Editing Techniques. *EMBO Reports* 17: 279–280.

Bourguignon, Didier. 2015. The Precautionary Principle. Definitions, Applications and Governance. http://www.europarl.europa.eu/RegData/etudes/IDAN/2015/573876/EPRS_IDA%282015%29573876_EN.pdf. Accessed 29 March 2018.
Church, Georg. 2017. Compelling Reasons for Repairing Human Germlines. *New England Journal of Medicine*: 377: 1909–1911.
Collins, Francis S. 2015. Statement on NIH Funding of Research Using Gene-editing Technologies in Human Embryos. https://www.nih.gov/about-nih/who-we-are/nih-director/statements/statement-nih-funding-research-using-gene-editing-technologies-human-embryos. Accessed 28 March 2018.
Council of Europe. 2017. 20th Anniversary of the Oviedo Convention. https://www.coe.int/en/web/bioethics/20th-anniversary-of-the-oviedo-convention. Accessed 29 March 2018.
Council of Europe. 1997. Explanatory Report to the Convention for the Protection of Human Rights and Dignity of the Human Being with Regard to the Application of Biology and Medicine: Convention on Human Rights and Biomedicine. https://rm.coe.int/16800ccde5. Accessed 28 March 2018.
Cwik, Bryan. 2017. Designing Ethical Trials of Germline Gene Editing. *New England Journal of Medicine* 377: 1911–1913.
Dance, Amber. 2017. Better Beings? *Nature Biotechnology* 35: 1006–1011.
Egli, Dieter, et al. 2017. Inter-homologue Repair in Fertilized Human Eggs? *bioRxiv*. doi:10.1101/181255.
ETC Group. 2016. Stop the Bomb! ETC Group Comments on NAS Report on Gene Drives. http://www.etcgroup.org/sites/www.etcgroup.org/files/files/etc_genedrivesnr_20160608.pdf. Accessed 29 March 2018.
European Commission. 2017. Special Eurobarometer 461. Designing Europe's Future. Trust in Institutions; Globalisation; Support for the Euro, Opinions about Free Trade and SIlodarity. http://ec.europa.eu/commfrontoffice/publicopinion/index.cfm/ResultDoc/download/DocumentKy/78720. Accessed 29 March 2018.
European Commission. 2016. Standard Eurobarometer 86. http://data.europa.eu/euodp/en/data/dataset/S2137_86_2_STD86_ENG. Accessed 29 March 2018.
Feinberg, Joel. 1980. The Child's Right to an Open Future. In *Whose Child? Children's Rights, Parental Authority, and State Power.* Aiken, William (ed.), 124–153. Totowa: Rowman & Littlefield.
Friedmann, Theodore, et al. 2015 ASGCT and JSGT Joint Position Statement on Human Genomic Editing. *Molecular Therapy* 23: 1282.
Gaskell, George, et al. 2017. Public Views on Gene Editing and Its Uses. *Nature Biotechnology* 35: 1021–1023.
German Ethics Council. 2017a. Annual Meetings. http://www.ethikrat.org/events/annual-meetings. Accessed 29 March 2017.
German Ethics Council. 2017b. Germline Intervention in the Human Embryo. German Ethics Council Calls for Global Political Debate and International Regulation. http://www.ethikrat.org/publications/ad-hoc-recommendations/files/recommendation-germline-intervention-in-the-human-embryo.pdf. Accessed 28 March 2018.
Huntington, Samuel P. 2003. *Clash of Civilizations and the Remaking of World Order.* London: Simon & Schuster.
Hyun, Insoo, and Catherine Osborn. 2017. Query the Merits of Embryo Editing for Reproductive Research Now. *Nature Biotechnology* 35: 1023–1025.

Jinek, Martin, et al. 2012. A Programmable Dual-RNA-Guided DNA Endonuclease in Adaptive Bacterial Immunity. *Science* 337: 816–821.

Kang, Xiangjin, et al. 2016. Introducing Precise Genetic Modifications into Human 3PN Embryos by CRISPR/Cas9-Mediated Genome Editing. *Journal of Assisted Reproduction and Genetics* 33: 581–588.

Kornwachs, Klaus. 2013. *Philosophie der Technik*. München: C.H. Beck.

Lanphier, Edward, et al. 2015. Don't Edit the Human Germ Line. *Nature* 519: 410–411.

Liang, Puping, et al. 2015. CRISPR/Cas9-Mediated Gene Editing in Human Tripronuclear Zygotes. *Protein & Cell* 6: 363–372.

Luhmann, Niklas. 1996. *Protest. Systemtheorie und soziale Bewegungen*. Frankfurt: Suhrkamp.

Luhmann, Niklas. 1995. *Social Systems*. Stanford: Stanford University Press.

Luhmann, Niklas. 1990. *Paradigm Lost. Über die ethische Reflexion der Moral. Rede anläßlich der Verleihung des Hegel-Preises 1989*. Frankfurt: Suhrkamp.

Ma, Hong, et al. 2017. Correction of a Pathogenic Gene Mutation in Human Embryos. *Nature* 548: 413–419.

Pew Research Center. 2015. Americans, Politics Science Issues. http://assets.pewresearch.org/wp-content/uploads/sites/14/2015/07/2015-07-01_science-and-politics_FINAL-1.pdf. Accessed 29 March 2018.

Rau, Johannes. 2001. Berliner Rede 2001 von Bundespräsident Johannes Rau. http://www.bundespraesident.de/SharedDocs/Reden/DE/Johannes-Rau/Reden/2001/05/20010518_Rede.html. Accessed 28 March 2018.

Ried, Jens, et al. 2011. Unbehagen und kulturelles Gedächtnis. Beobachtungen zur gesellschaftlichen Deutungsunsicherheit gegenüber Synthetischer Biologie. In *Was ist Leben – Im Zeitalter seiner technischen Machbarkeit. Beiträge zur Ethik der Synthetischen Biologie*. Dabrock, Peter et al. (eds.), 345–369. Freiburg: Karl Alber.

Starkbaum, Johannes, et al. 2015. The Synthetic Biology Puzzle: A Qualitative Study on Public Reflections towards a Governance Framework. *Systems and Synthetic Biology* 9: 147–157.

The Academy of Medical Sciences, et al. 2015. Genome Editing in Human Cells – Initial Joint Statement. https://acmedsci.ac.uk/file-download/37773-55e6b4e90f49c.pdf. Accessed 28 March 2018.

The Economist. 2015. The Age of the Red Pen. *The Economist*, August 22.

The Hinxton Group. 2015. Statement on Genome Editing Technologies and Human Germline Genetic Modification. http://www.hinxtongroup.org/hinxton2015_statement.pdf. Accessed 28 March 2018.

The National Academies of Sciences, Engineering, and Medicine. 2017. *Human Genome Editing. Science, Ethics and Governance*. Washington, DC: The National Academies Press.

The National Academies of Science, Engineering, and Medicine. 2015. On Human Gene Editing: International Summit Statement. http://www8.nationalacademies.org/onpinews/newsitem.aspx?RecordID=12032015a. Accessed 28 March 2018.

Thomas, William Isaac, and Dorothy Swaine Thomas. 1928. *The Child in America. Behavior Problems and Programs*. New York: Knopf.

United Nations Education, Scientific and Cultural Organisation (UNESCO). 1997. Universal Declaration on the Human Genome and Human Rights. http://portal.unesco.org/en/ev.php-URL_ID=13177&URL_DO=DO_TOPIC&URL_SECTION=201.html. Accessed 28 March 2018.

United Nations Education, Scientific and Cultural Organisation (UNESCO); International Bioethics Comittee (IBC). 2015. Report of the IBC on Updating Its Reflections on the Human

Genome and Human Rights. http://unesdoc.unesco.org/images/0023/002332/233258E.pdf. Accessed 29 March 2018.
Wadhwa, Vivek. 2015. Why There's an Urgent Need for a Moratorium on Gene Editing. *The Washington Post*, September 8.
Wissenschaft im Dialog. 2017. Wissenschaftsbarometer 2017. https://www.wissenschaft-im-dialog.de/medien/publikationen-fotos-videos/wissenschaftsbarometer-2017/. Accessed 29 March 2018.
Wolpe, Paul Root, et al. 2017. Ethical Principles for the Use of Human Cellular Biotechnologies. *Nature Biotechnology* 35: 1050–1058.

Debating Genome Editing Technologies
Questioning the Role of Public Discussion

Lukas Kaelin

Keywords

Public Opinion, Public Sphere, Political Theory, Deliberative Democracy

Abstract

Ethical statements and position papers on genome editing often refer to "the public" that should get involved, and a public discussion that should take place. It is not self-evident what this reference to the public means and why such a public engagement should take place. This chapter explores the concept of the public as discussed in key position papers on genome editing and puts it into relationship with the notion of the public in political theory. The fuzzy notion of the public is presented in these statements on genome editing as a tool to solve problems and to allow for hitherto unheard points of view to be considered. It is sometimes understood as a general tool for reciprocal education and social self-determination. Other times still it is used in an instrumental way to gain public support and foster an environment friendly to research and innovation. There is disagreement about the aims, the reason, and the norms guiding public discussion and the possibility of a neutral form of communication. Last but not least awareness is required for the way the public debate is framed by the very use of "genome" and "editing" rather than other concepts.

M. Braun et al. (Eds.), *Between Moral Hazard and Legal Uncertainty*,
Technikzukünfte, Wissenschaft und Gesellschaft / Futures of Technology,
Science and Society, https://doi.org/10.1007/978-3-658-22660-2_12

1 Introduction

Ethical statements on new genome editing technologies, such as CRISPR-Cas9, often stress the importance of public discussion and the participation of civil society in the development of the norms that guide this research and the formulation of sustainable regulation. The European Group on Ethics in Science and New Technologies "call[ed] for a broad public debate on these issues" (EGE 2016), the Berlin-Brandenburg Academy of Sciences and Humanities demanded "a carefully moderated, responsible and differentiated ethical discussion" (BBASH 2015, 8), and the National Advisory Commission on Biomedical Ethics stressed the need for an "intensive, critical, open, and transparent social debate" and wanted to contribute to "a public discussion by organizing public events" (NCE 2016, 4). More recently an editorial in *Nature* argued not only for further research groundwork and an examination of alternatives, but also for "further societal debate" before clinical applications can be considered (Nature 2017).

In view of the potential benefits and risks as well as the encompassing expectations and fears that accompany the development of genome editing, there is a widespread belief in the scientific community, among research organizations and other stakeholders, that the broader public should be engaged in the development of these technologies. This chapter aims to trace the different understandings of the public's role in debates about genome editing, putting them in relation to political theories of the public sphere. In order to provide such an account, I focus mainly on the position papers published by key academic institutions and analyze their understanding of the public. In my approach to tackle the question of the role of the public, I also outline the basic concept of the public sphere. This is an important step, since there is considerable disagreement among political philosophers and theorists in the understanding of the public and its function. The different aspects of this fuzzy notion are then linked to the way the public is discussed in the different position papers on genome editing.

2 The Public Sphere as Collective Problem Solving

Most generally, the public sphere can be understood as the medium in which the society becomes conscious of itself. In this most general sense the public sphere is – as Volker Gerhardt puts it – the "political form of consciousness" (Gerhardt 2012), and thus, a basic dimension of human social and political life. The public is the necessary flipside of domestic life; it is the social and political aspect of human existence. Put emphatically, it is the place where human life in its essential sense takes place. Hannah Arendt's concept of the public as developed from ancient Greece fits into this idea of the existential dimension of public life. In a functional analysis, one might stress the self-organization that can take place in such a sphere: As a community we gather as a public to lay down the rules that govern our social life. Such a conceptualization of the public sphere, however, is too broad to serve as an analytical tool for explaining political public discourse. It is also too broad to analyze the role of the public in the debate on genome editing.

If we understand the public in less emphatic and more pragmatic terms, we might reach a better understanding of how it is used in the debate on genome editing. The nucleus of the public can be conceived as collective problem solving for issues that arise from social cooperation. Resulting from indirect consequences of cooperation among private individuals, a public is formed by affected third parties. As John Dewey famously states, "[t]he public consists of all those who are affected by the indirect consequences of transactions to such an extent that it is deemed necessary to have those consequences systematically cared for" (Dewey 2012, 48). This implies that the public sphere has a primarily reactive character. Problems, discontents, sufferings etc. bring about public debate and discussion for the purpose of problem-solving. Thus, it is vital for the public sphere to recognize these problems and develop the competences to bring about the necessary change. The main challenge of the public – as Dewey puts it – is of an intellectual nature (cf. Dewey 2012, 110). The task at hand is to properly analyze the relevant actions and the consequences that cause problems.

Crucial for this aspect of public discussion is the notion of being affected by the cooperation of others. But what does "being affected" mean? Questions of medical ethics and bioethics at times bring about a strong sense of moral feeling in a large segment of the population, even though the technology at stake has no (or little) impact on their lives. Embryonic stem cell research or preimplantation genetic diagnosis (PGD) are two such examples. Other issues create less social involvement, even though they affect large segments of the population. The fuzzy notion of "being affected" surfaces in position statements with the reference to "stakeholders". The National Academy (United States), for example, stresses the importance of engaging stakeholders in order to "increase public perceptions of the legitimacy of regulatory or policy decisions surrounding emerging technologies." (National Academies 2017, 164)

The notion of "being affected" cannot be reduced to any quantifiable economic or social impact, but has to be understood pragmatically. Public debates on bioethical issues arise where a number of individuals identify a problem and feel the need to act on it. There is a number of reasons why, following Dewey, the public is an intellectual problem especially in bioethical matters. First, many bioethical questions require cognitive capacities to grasp the issues. Public debates thus depend on the formation and education of citizens (see below), which might take place in the form of (a) quality scientific journalism, (b) a general level of good education and (c) a group of lay specialists who develop expertise in a particular social field (Peters 2007, 97–99). It is also an intellectual problem, because, second, knowledge is required to properly identify the relevant cooperation that brings about these consequences. Finally, third, the intellectual challenge of technological progress in the field of biotechnology requires knowledge about how proper regulation can be achieved given the international resp. global aspect of this research.

In the case of genome editing in this early stage of the technology, the "affected" public consists mainly of researchers, ethicists and the corresponding research organizations. They make up, so to speak, the core public of affected people, who in turn try to engage the larger public, as can be seen in the various statements and position papers. Yet, the perspective of the public as problem-solving gets an important twist. In the words of a memorandum of the Obama administration on emerging technologies, involving and engaging the public is seen as instrumental "for promoting accountability, for improving decisions, for increasing trust, and for ensuring that officials have access to widely dispersed information" (Holdren et al. 2011, 2). The reason there should be public engagement is argued by turning Dewey on his head: It is not a bottom-up approach, where affected parties try to regulate the technological development they have become aware of. Rather, the participants of direct cooperation (the researchers) are seeking the consent of those subjected to the indirect consequences of their actions (the larger public).

Such an instrumental understanding of the public can be detected, for example, in the National Academy's chapter on "public engagement". The National Academy's study gives different reasons for the necessity of engaging with the public. Engaging the public helps anticipate potential reactions, thus providing planning reliability. Failing to include the public in the introduction of controversial technologies might create an environment hostile to research, leading to the inhibition of innovation, stigmatization of technologies, and trade barriers (cf. The National Academies 2017, 164). Thus, the authors understand the reason for public engagement instrumentally when they argue that the "lack of meaningful engagement with different publics when genetically modified organisms (GMOs) were first introduced did irreparable damage to the emerging scientific field of genetic engineering" (The National Academies 2017, 164). Public engagement, then, is understood as a way to sensitize the public and to be able to react to its sentiment, in order to be more in control of the process of introducing new technologies. The public attitude towards genetically modified food is portrayed as detrimental to important global health tasks, like reducing vitamin A deficiency through the introduction of golden rice.

Stressing the importance of public engagement, to which I will return at the end of this chapter, has also to be understood in the context of the public's critical stance towards genetic engineering in GMOs. With regards to genome editing, there are calls for ongoing public, and not just scientific, discussions about key differences between disease and enhancement, between health and disability, which need to be publicly discussed the more genome editing is applied in practice (cf. The National Academies 2017, 164). Public engagement is not only advocated before the introduction of a new technology, rather the public should be involved continuously as different applications become possible.

3 Equal Participation as a Key Norm of the Public

In its ideal form, the public sphere can be understood as the discursive, (but not necessarily) argumentative solving of problems by everyone affected. The normative demand of "equality" – inherent in democratic forms of governance – can refer to an equal, actual say of everyone involved, or to the mere potential of voicing one's concern. In most public sphere theory "equality" is translated by some procedures of discussion that would fairly distribute the chances of participation. The ability to collectively solve these problems is, obviously, dependent on the (material and ideal) structure of the public. This dependency on the "media" (in the broad sense) composition of the public sphere encompasses social, political, historical, economical, mass media aspects. For analytical purposes, the discussion of the public sphere is normally reduced to the question of the mass media. However, it is important to emphasize these other aspects that allow (or do not allow) effective collective problem solving.

How this demand for "equality" is to be understood is subject to considerable disagreement. A closer look at the *International Summit on Human Gene Editing*, which was held in December 2015 in Washington and can be identified as the starting point of the (global) debate on genome editing, illustrates how the norms guiding the public debate are imagined and what sort of equality is emphasized. Hosted by the academies of science (and medicine) of the United States and China as well as United Kingdom's Royal Society, many participants at the conference stressed the importance of a discussion involving participants from different cultures around the world as well as the broader (lay) public. The prominence of this meeting consists in its global scope and the explicit concern, throughout the conference, to engage the public as well as the position of the conference right at the beginning of the debate about genome editing. It therefore provides first considerations and reflections – the first raw material in the discussion, so to speak – that have been further substantiated in the course of the debate.

Among the participants, and of the commissioned papers, David Baltimore's contribution puts the strongest emphasis on the public. He emphasizes the need for a global discussion that includes "the voices of diverse cultures" and (after mentioning different scientific communities) "of course the lay public" (International Summit 2015, 4). He sees the conference as the starting point for such a discussion. It remains ambivalent whether such a discussion should be just one-way, conveying knowledge to a yet to be informed public, or whether the role of the public is seen as more interactive. In his view, the public (in a very general, undifferentiated sense) "has become ever more aware of the power of science"; it "has witnessed the huge benefits of basic and medical research" and "has become more engaged in debates about science and scientific progress" (International Summit 2015, 5). But why do we need to engage the public? Baltimore gives no reason. Rather, he postulates the need for a broad discussion. The information of the public seems to be instrumental for the acceptance of any research, in order to address the general public who is perceived as being worried about biotechnological research getting out of control. Equality is understood in the form of an engagement of different cultures and the inclusion of scientists and a lay public alike.

Yet, there is a need for further reflection on the meaning of equality and distinguishing it from equity. Charis Thompson raises the question about the "missing debates and constituencies at this meeting" (International Summit 2015, 46). She lists a number of perspectives (disability, race, gender, queer bioethics, health disparities, commercialization, medical tourism, bioart and biohacking, biosecurity, other species) that need to be considered, but are often marginalized in the debate. Public participation, therefore, means that all these different perspectives can make their voices heard in the debate about genome editing. Thompson's idea of the public is akin to the Habermasian notion of discourse (see below), which requires that all arguments are heard and that no groups are excluded from the debate.

Ruha Benjamin emphasizes the disability perspective. In her research, done in California, on stem cell research, she observed that patient's advocates were included to make the research more accountable to the people. Yet, these patient advocates "framed their demands in terms of medical consumer rights" (International Summit 2015, 49) and pursued an agenda most conducive to the middle classes. It is doubtful whether these patient advocates represent "disabled people for whom dismantling policies and prejudices that cast them as second class is often more vital than access to 'miracle cures'" (International Summit 2015, 49). She calls the particular type of public represented by patient advocates the "default public." Her paper serves as a reminder that the mere invocation of the public, without any further specification (or even open access to interested parties), can produce problematic effects for minorities who are unable to enter the debate.

Different nations have different sensitivities concerning the danger of excluding certain groups, and they might take diverse measures to ensure that minorities' views are considered. The disability perspective that Ruha Benjamin stressed, for example, was long absent from the global discussion on genome editing and might still be ignored in certain discourses. The difference in the awareness for ensuring that different population groups are heard in the discussion can be seen in the composition of the different national ethics councils. Belgium, for example, ensures that the main languages are equally represented; France puts emphasis on the representation of different religious and political ideologies; in the United States particular emphasis is put on regional, gender and racial participation (Fuchs 2005).

4 The Education of the Public

In the very general sense, the public sphere is a universal formation process ("Bildungsprozess"); people form opinions on matters they deem relevant. How well-informed these opinions are may differ significantly, but even ill-informed opinions have to be developed by gaining information about social reality. On a normative plane, the quality of the public sphere depends on the education/formation of the participants engaging each other in the public sphere. In an ideal sense, the public sphere has a formative (in the normative sense), i. e., emancipative character, allowing participants to better understand their reality and thereby – in a Kantian sense – allowing them to pass from a state of immaturity to one of maturity ("Mündigkeit").

Such an understanding of the public is detectable in Ismail Serageldin's opening statement to the Washington conference. He stresses the importance of "ethics and public understanding […] to help our societies [to] better cope with the rapidly changing technological scene" (International Summit 2015, 1–2). The idea here seems to be that an informed public is better equipped to "cope" with new technologies. However, it is not prima facie evident why a society needs such understanding. Looking to other technologies, there is ample evidence that societies can use and apply them without a broad understanding by the larger public. Put in the broader context, one might understand this evocation of the "public" as a quite general claim of an enlightened citizenry in the tradition of Kant's appeal to the "public use of one's reason in all matters" (Kant 1784). Kant's understanding of the public requires all citizens to make use of their reason and share their understanding of any given matter with each other. The "public use of reason" is instrumental for Kant in his emphatic understanding of enlightenment; it is by virtue of reciprocal information and discussion that citizens are able to leave their state of immaturity.

In its broadest sense, the call for the involvement and information of the public in bioethical questions aspires to a situation in which citizens are no longer ignorant about the structures of their lives and their society, but are enabled to understand, criticize, and eventually change the structures to which they are subjected. In Serageldin's vague opening statement, however, the purpose of the "public understanding" has to be taken more instrumentally in terms of society's coping mechanisms with reference to rapid technological developments.

This very general and unspecific evocation of the public is present in many position papers. The statement made by the National Advisory Commission on Biomedical Ethics (NCE) of Switzerland, for example, basically presents the different arguments and positions of members of the NCE without taking a position. The short statement sketches the different positions, in the discussion on genome editing, that have already been put forward in other discussions on the status of the embryo, the danger of instrumentalization of human life, human experimentation, and social ethical constraints. The concluding remarks emphasize the importance of an "intensive, critical, open and transparent societal debate" (NCE 2016). The commission is in favor of a legal, normative discussion about the scope of permission or prohibition of germline manipulation. Yet, it does not argue the reasons for the need of a public discussion. As all Swiss national laws are subject to an optional referendum (with a comparably low threshold of signatures required), there is a known option for politically organized stakeholders to force a legally binding referendum.

5 Between Open Discourse and Struggle for Hegemony

The public sphere can be described empirically as a field in which different players compete for attention and influence. The garnered attention can be used in different ways to gain influence, but attention does not automatically translate into political influence. Ultimately, this sociological perspective on the public sphere cannot be properly reconciled with the core of an emphatic normative model, as proposed by Bernhard Peters (2007) and Jürgen Habermas (1996), that aims at mutual understanding among agents in the public sphere. Theoretical strategies for dealing with this tension consist in either emphasizing the political nature of the public sphere, in terms of struggle for hegemony (Chantal Mouffe), or in focusing on the implicit normative self-understanding of the public (Jürgen Habermas).

With regard to the notion of the public and the role it is supposed to play, the Nuffield Council articulates a quite particular understanding of the way bioethical debates unfold (and should unfold). Its understanding of the public is somehow situated between an open discourse and a struggle for hegemony. In a dense paragraph it describes the way expectations are structured through the presentation of the prospects of a particular technology, that might lead to the pursuit of these technologies. In turn, other technologies remain unexplored and unrealized, as attention, funding, and research are shifted towards other options. This leads, according to the argumentation of the Nuffield Council, to "over-claiming in areas in which science encounters politics" (Nuffield 2016, 19). There is an increasing awareness of this problem and of the need to present research options as candidly and as soberly as possible. However, innovation requires these "visions of desirable and scientifically attainable futures" (Nuffield 2016, 19): "It is not necessary (or possible) that this should take place in neutral language or against a background of acknowledged priorities and values; what is dangerous is where there are asymmetries of power, information or representation in the public sphere that mean that certain visions and values go unappreciated and others go unchallenged" (Nuffield 2016, 19).

The last sentence raises important questions, which are at the heart of public sphere theory. On the one hand, the tradition following Jürgen Habermas places a strong emphasis on rational debate, where all arguments are heard and no one is excluded: His model puts emphasis on a neutral ground (and neutral language), where the discussion takes place. Crucial for public discussions is "a *shared* practice of communication […]. Agreement on issues and contributions develops only as the result of more or less exhaustive controversy in which proposals, information, and reasons can be more or less rationally dealt with" (Habermas 1996, 362). The quality of the outcome of public discussion is thus dependent on all arguments being taken into consideration and on the rationality of the debate. In his discussion on religion Habermas seeks to clarify the conditions for public debate, where he stresses the need for translation from ideologically framed beliefs into a neutral language (Kaelin 2008).

This model, however, is challenged, for example, by Chantal Mouffe. Any public discussion is a result of the hegemonic configuration of society. Not all issues are negotiable in public; rather, there are taboos, ideological constraints, and different forms of exclusion. Stressing the need for a "rational" debate presupposes a neutral ground that does not exist, and is itself an expression of the hegemonic constitution of the contemporary public sphere (e.g. Mouffe 2013). Mouffe views the public sphere as a place of constant struggle for which the notion of opponents is crucial. Debating in the public implies conflict over ways of life and deep-seated values. Private beliefs about personal ways of life cannot be excluded from public deliberation. And the question of what can be subject to public discussion is the result of the existing powers in society, i. e., the way (and likelihood) of discussing a particular topic (in our case genome editing) is a result of societal powers.

Every such discussion is shaped by discursive power that includes some people and excludes others. Alexander Bogner, who analyzed citizen conferences on bioethical topics in Germany, describes these carefully planned forms of participation of the larger public. They have led not only to the marginalization of deliberation, but also to many participants dropping out during the process, as they could not "deal with" the mode of participation – because they could not accept pluralism of opinion, because they had strong/unusual beliefs, or because they felt marginalized (cf. Bogner 2011, 174–175). Findings from these citizen conferences may be adapted to the discussion in the larger public.

In light of this disagreement about the proper understanding of the public sphere – exemplified by the positions of Habermas and Mouffe – the Nuffield Council's statement tries to strike a balance. It acknowledges the impossibility of a neutral language, yet it still upholds the vision of a basis of discussion that is without the asymmetries of power that mute certain arguments. At the same time it glosses over important rifts in public sphere theory and avoids important conceptual questions about their understandings of "power asymmetries" or "appreciation of all values."

The Nuffield paper further points out that including the public in the procedure about how to proceed with new technologies, such as genome editing, is a vital part of democratic governance. Although democratic procedures are "imperfect, slow, difficult and expensive", they are seen by the Nuffield Council as offering "a plausible solution to, or way of coping with, the problem of the mutual adaptation of emerging biotechnologies and the normative frameworks within which they are deployed" (Nuffield 2016, 31). The Nuffield Council closes this chapter with the sentence: "Much of the evidence we received pointed to the importance of having an open, effective and inclusive public sphere in which questions about genome editing could be raised and discussed, in which different positions and arguments could encounter each other, and the importance of democratic governance." (Nuffield 2016, 13)

In preparation for their report the Nuffield Council issued a "Call for evidence" (cf. Nuffield 2016, 125) that asked organizations and individuals to answer questions about their hopes and concerns regarding genome editing. Their stress on the public sphere seems to be a direct consequence of the responses they got. This might explain the contradictory notion of this paragraph, which glosses over important issues in the development of genome editing. Democratic governance via the public sphere is a general request from many interested individuals and organizations, yet it slows down technological progress in a very competitive field. Pointing out the downsides of democratic governance hints at other forms of the governance of technology better equipped to foster research and innovation. The engagement of the public is depicted as a delaying element in innovation, yet a necessary one for a technology as sensitive in the public perspective as genome editing.

6 The Framing of the Debate

One important aspect of the concept of the public, related to its ambivalent status between open discourse and the struggle for hegemony, is the way a debate is framed. This question has recently gained attention in the field of politics, where the work of George Lakoff, among others, stresses the way public debates are preconditioned by the concepts used. This point is raised by the elaborated paper, published by the Nuffield Council, when discussing "genome editing" as a metaphor, and what that means for the entire debate. "The 'editing' metaphor distinguishes the approach from less 'precise' forms of genetic 'engineering'" (Nuffield 2016, 19). But what does "more precise" mean? "Precision" has to be understood as a thick concept with normative overtones, hinting at controlled intentionality (i. e., in "precision medicine" or "precision warfare") and needing differentiation. Speaking of "editing" (instead e. g. of engineering) implies an "editor", who is in charge of the editing and who is in control of the process in a more clinical way than in the case of "engineering." These reflections on genome editing as a metaphor bring to consciousness how our reflections on technologies are framed by the concepts used. If we follow the research of George Lakoff, then metaphors are not ways merely of understanding the world, but of the very way we think (Lakoff et al. 2003). The use of a particular metaphor, therefore, shapes the following discussion and is relevant for the normative judgments that might result.

This general reflection on genome editing as a metaphor has to be understood in the context of the promises and fears that tend to accompany emerging technologies. Genome editing has to be understood in the manner of emerging technologies, which are, as the Nuffield Council points out, "promissory by nature, [and therefore] characteristically subject to 'hype' and over-claiming" (Nuffield 2016, 19). Any discussion about new technologies will have this characteristic, informed by speculation and excessive expectations, for positive as well as negative consequences.

This reflection on the framing of the debate echoes Charis Thompson's statement at the Washington conference, which focuses on public participation. She raises two important points, which remained absent from the other talks. Her first point is about the metaphors used in the debate and, therefore, how the issues are framed. She raises questions such as: Do we use "gene" or "genome", and how does that impact our value judgments? What do we associate with "editing"? Do we talk about "adult" or "somatic" stem cells? Her point is not so much to favor any particular way of talking about the issue, but rather to raise the awareness that these questions matter to the public.

7 Searching for Public Engagement and Consent

Throughout the different statements and position papers, there is a recurring theme for the engagement of the public. This theme, already touched on at the beginning in the context of problem-solving, is the search for public consent to and approval of the research of genome editing, to gain public funding and create social acceptance for developed products. This search for consent is present, for example, in Alta Charo's statement about legal and regulatory context. She stresses the importance of "consumer demand" for the development of the particular products and services of genome editing. While public comment is a fairly soft regulatory instrument, it nevertheless "creates political pressure that can drive decisions in one way or another, and it allows for some interplay between government expertise/authority and public consultation" (International Summit 2015, 14). In this notion of public comment or consultation, a regulatory process is opened for intervention, which will steer regulation through by means of political pressure. The public in Charo's statement is important, on the one hand, because of the consumer power, which can promote particular products and services. Ignoring the public would be detrimental for anyone pursuing a particular research. The research might need public funding; its applications will need public interest. On the other hand, public consultation can direct research in certain ways rather than others.

Yet, there is substantial disagreement about what public engagement consists in. The study by the National Academy identifies three principles of public engagement: (a) quality of outcomes, (b) legitimacy of outcomes, and (c) administrative efficiency. Public engagement procedures can improve the quality of policy decisions (a) in four distinct yet related ways. First, public engagement allows for considering "systematically the widest possible range of effects, as well as uncertainties surrounding them" (The National Academies 2017, 166). Second, all regulatory and policy options can come into consideration, too. Third, public engagement allows for better understanding of the way technologies might change the values of the people affected. Finally, questions, ideas and solutions previously not considered, might be brought into the discussion by members of the lay public.

The legitimacy of outcomes (b) is increased, if the process of decision making is transparent and "perceived by all parties as fair and competent" (The National Academies 2017, 166). Involving all interested and affected parties is important for the legitimacy of the process. These two principles have to be balanced with the third one, administrative efficiency (c). As the National Academy writes with regard to genetically engineered crops: "The goal of full participation needs to be considered in light of the need for administrative efficiency to ensure that decisions are made in a timely manner" (The National Academies 2016, 55–56).

This depiction of the three principles of public engagement remains slightly fuzzy and raises important questions. Obviously the way "efficiency" is balanced against "inclusion" needs a discussion about procedures by which such a balancing can work in practice. Inclusion of the public is, to quote the Nuffield Council, "imperfect, slow, difficult and expensive" (Nuffield 2016, 31). The first two principles aiming at a high quality of the outcomes and legitimacy provide an ideal that can only be partially realized. Democratic elements tend to slow down administrative efficiency.

But another tension has to be added to the one between these three principles. The report glosses over the inherent problems of the notion of legitimacy as described in the discussion on deliberative democracy. There is a tension between participation and publicity, on the one hand, and the quality of deliberation on the other hand. Broad participation lowers the quality of deliberative reasoning, as Dennis Thompson (2008, 513) points out. This tension might be constructed as a trilemma between political equality, mass-participation, and deliberation (Fishkin 2009, 191): [P]olitical equality and mass-participation lead to a direct democracy with low quality of deliberation. Political equality and deliberative quality miss out on mass-participation. "Achieving deliberation and participation can be achieved for those unequally motivated and interested, but violates political equality." (Fishkin 2009, 191)

8 Summary

This chapter has examined the different ways the public is used in position papers and statements on genome editing and put it in relation to different theoretical accounts of public sphere theory. Given the fuzzy nature of the concept of the public, it is not surprising that the public is mentioned in different ways and supposed to fulfill different functions. The public is sometimes understood in its ability to foresee and solve problems emerging from new technologies, sometimes it is simply engaged in order to create an atmosphere friendly to research and innovation. At other times the public is understood in terms of general enlightenment ideals to allow citizens to decide about technologies they will be subjected to. When it comes to the way the debate is supposed to take place, there is a general disagreement in public sphere theory. The demand for a neutral ground and an objective and rational debate is countered with the claim that all public debates essentially are framed by social and political configurations of hegemony. Trying to leave out passions and the entire private life would not only not be feasible, it would also not be desirable. The supposed role of the public in the debate on genome editing reflects that disagreement.

The last two points of this chapter stress the different aspects of the discussion of the public in the different position papers on genome editing. The way the debate is already framed by using the terms "genome" (rather than "gene") and "editing" (rather than "engineering") already sets the debate in a nexus of concepts and perceptions, e. g., of precision, and thereby shapes the debate in advance. The last point, on the search for public engagement and consent, highlights once more that the debate on genome editing is instrumental in as far as it tries to create in the population an attitude of openness to innovation and research.

References

Academy of Sciences Leopoldina. 2017. Ethical and Legal Assessment of Genome Editing in Research on Human Cells. https://www.leopoldina.org/uploads/tx_leopublication/2017_Diskussionspapier_GenomeEditing.pdf. Accessed 7 December 2017.

Baltimore, David, et al. 2015. On Human Gene Editing: International Summit Statement. http://www8.nationalacademies.org/onpinews/newsitem.aspx?RecordID=12032015a. Accessed 19 January 2018.

Berlin-Brandenburg Academy of Sciences and Humanities (BBASH). 2015. Human Genome Surgery – Towards a Responsible Evaluation of a New Technology. http://www.gentechnologiebericht.de/bilder/BBAW_Human-Genome-Surgery_PDF-A1b-1.pdf. Accessed 7 December 2017.

Bogner, Alexander. 2011. *Die Ethisierung von Technikkonflikten. Studien zum Geltungswandel des Dissenses*. Göttingen: Velbrück.

Dewey, John. 2012. *The Public and Its Problems*. Pennsylvania: Pennsylvania State University Press.

European Group on Ethics in Science and New Technologies. 2016. Statement on Gene Editing. https://ec.europa.eu/research/ege/pdf/gene_editing_ege_statement.pdf#view=fit&pagemode=none. Accessed 7 December 2017.

Fishkin, James S. 2009. *When the People Speak*. Oxford: Oxford University Press.
Fuchs, Michael. 2005. *Nationale Ethikräte. Hintergründe, Funktionen und Arbeitsweisen im Vergleich*. Berlin: Saladruck.
Gerhardt, Volker. 2012. Öffentlichkeit. Die politische Form des Bewusstseins. München: C.H. Beck.
Habermas, Jürgen. 1991. *The Structural Transformation of the Public Sphere*. Cambridge, Mass: MIT Press.
Habermas, Jürgen. 1996. *Between Facts and Norms*. Cambridge: MIT Press.
Holdren, John, et al. 2011. Memorandum: Principles for Regulation and Oversight of Emerging Technologies. https://obamawhitehouse.archives.gov/sites/default/files/omb/inforeg/for-agencies/Principles-for-Regulation-and-Oversight-of-Emerging-Technologies-new.pdf. Accessed 15 February 2018.
International Summit on Human Gene Editing. 2015. A Global Discussion. http://nationalacademies.org/cs/groups/pgasite/documents/webpage/pga_170455.pdf. Accessed 8 December 2017.
Kaelin, Lukas. 2008. Habermas, the Jesuits and Religion: Notes on a Discussion about the Role of Religion in Society. *Prajñā Vihāra* 9: 15–30.
Kant, Immanuel. 1784. What is Enlightenment? http://www.columbia.edu/acis/ets/CCREAD/etscc/kant.html. Accessed 15 February 2017.
Lakoff, George, and Mark Johnson. 2003. *Metaphors We Live By*. Chicago/London: University of Chicago Press.
Lippmann, Walter. 2010. *Public Opinion*. Miami: BN Publishing.
National Advisory Commission on Biomedical Ethics (NCE). 2016. *Gene editing an menschlichen Embryonen – Eine Auslegeordnung*. http://www.nek-cne.ch/fileadmin/nek-cne-dateien/Themen/Stellungnahmen/NEK_Gene_editing_Papier_web_DEF.pdf. Accessed 7 December 2017.
The National Academies of Sciences, Engineering, and Medicine. 2016. *Genetically Engineered Crops: Experiences and Prospects*. Washington, DC: The National Academies Press.
The National Academies of Sciences, Engineering, and Medicine. 2017. *Human Genome Editing: Science, Ethics, and Governance*. Washington, DC: The National Academies Press.
Nature Editorial. 2017. Ethical embryo editing. *Nature* 549: 307.
Mouffe, Chantal. 2013. *Agonistics. Thinking the World Politically*. London/New York: Verso.
Peters, Bernhard. 2007. *Der Sinn von Öffentlichkeit*. Frankfurt: Suhrkamp.
Sunstein, Cass. 2007. *Republic.com 2.0*. Princeton/Oxford: Princeton University Press.
Thompson, Dennis. 2008. Deliberative Democratic Theory and Empirical Political Science. *Annual Review of Political Science* 11: 497–520.

A Path Through the (Moral) Morass

Genome Editing, Reproduction and Broad Conversations[1]

Giulia Cavaliere

Keywords

Assisted Reproduction, Deliberative Democracy, Moral Judgment, Pluralism, Safety

Abstract

In this chapter, I discuss the possibility of using genome editing technologies in the context of assisted reproduction. I present the most prominent arguments in favour of and against this use of genome editing technologies, and explore two strategies used in bioethics in the event of moral disagreement to analyse the questions at hand and to develop morally sound policies. These strategies are: the clarification of the facts regarding a new technology and the formulation of the best philosophical argument. I contend that both approaches are theoretically flawed and might result in states of affairs that fail to promote peoples' well-being. I then argue that we should focus instead on finding at least some common ground in order to move forward in the debate on genome editing, building a regulatory framework that lightens "the burden of human existence" (Harris 2016), whilst still accommodating opposing views as much as possible. Furthermore, I argue that insights from moral psychology and democratic governance can assist us in these challenges.

1 This paper is part of a Wellcome Trust Project (WT108623/Z/15/Z) titled "Preimplantation Genetic Diagnosis and Eugenics: A Social Moral Epistemology Approach" awarded to Giulia Cavaliere (October 2015-October 2018). The author would like to thank Silvia Camporesi, Lorenzo Del Savio, Alberto Giubilini, John Harris, Barbara Prainsack and Hannah Schickl for their valuable comments on earlier versions of this chapter.

M. Braun et al. (Eds.), *Between Moral Hazard and Legal Uncertainty*, Technikzukünfte, Wissenschaft und Gesellschaft / Futures of Technology, Science and Society, https://doi.org/10.1007/978-3-658-22660-2_13

1 Genetic Technologies and (Moral) Disagreement

New reproductive technologies and screening technologies allow couples who are at risk of transmitting genetic diseases to their offspring to reduce the risk of transmission while still being able to have genetically related (or partially genetically related) offspring. Most of these technologies and their applications are controversial. There is no consensus regarding the legitimacy of their use and whether or not they are a morally acceptable means in the pursuit of parenthood projects. Some people oppose pre-natal and pre-implantation screening technologies because of their discriminatory nature and because they express – or strengthen – negative attitudes towards people with disabilities (Parens and Asch 2003). Others contest that these technologies could bring about increased inequality as well as sexist and racist attitudes (Roberts 1997). Others criticise them for violating human dignity and for tampering with human nature (Kass 1997), while still others are more concerned with the reinforcement of certain beliefs about the importance of genetic parenthood and the traditional family (de Melo-Martín 2017a). If there were agreement regarding these criticisms, bioethicists would be out of a job (together with the scientists developing them, the healthcare professionals working in the fertility clinics, and the policy makers regulating their use).

Fortunately (I am a bioethicist and I do not want to be out of a job), these technologies are also welcomed by many, since they allow people to pursue their preferred life plans, to have children that are genetically related to them, and who are healthier than they otherwise would have been. They, as Harris (2016) puts it, help "lighten the burden of human existence" (Harris 2016, 16). In this chapter, I focus on genome editing technologies and on their applications, in the context of assisted reproduction, for the correction of genetic abnormalities in embryos created with in vitro fertilization (IVF). I discuss this possibility, presenting some insights from the literature on moral psychology which can be considered a starting point for a debate on the challenges of institutional design.

Gene editing technologies have been around for over a decade now (Nuffield Council on Bioethics 2016). Despite this fact, one new gene editing technique in particular sparked a heated debate early in 2015, and to date, the debate has yet to be settled. Calls for a moratorium on the use of this particular technique as well as for international bans and joint efforts to prevent its applications in the clinic abound (Baylis 2017a and 2017b; Baltimore et al. 2015; Lanphier and Urnov 2015; UNESCO IBC 2017). The technique in question is CRISPR (Clustered Regularly Interspaced Short Palindromic Repeat), an RNA-guided tool that makes use of a naturally occurring defence mechanism employed by bacteria to avoid harmful infections caused by pathogenic organisms (e.g. viruses). The RNA tool (CRISPR) functions as a guide for other systems (Cas9, CPF1 and others) to target specific parts of the genome, which are subsequently cut by the Cas proteins. These cut strands can be exploited to modify the nucleotide sequence of DNA and to insert DNA sequences at the cut site. The application of this technique to human embryos and human gametes (i. e., oocytes and sperm cells) has been widely criticised for a number of reasons, but chiefly for its potential to introduce inheritable changes in the human genome. The question of consensus has catalysed the attention of many scientists and ethicists and has transformed CRISPR into one of the most discussed technologies of the past years (Brokowski et al. 2015; Lander 2015; Lanphier and Urnov 2015). The technique's precision, effectiveness and relatively limited cost, together with its potential to edit the human germline in a targeted manner, which elevates it above many other genome editing technologies, have led ethicists, scientists, reporters and members of the public to call for international solutions to what is perceived to be an international challenge (Baylis 2017a and 2017b; The National Academies 2015).

2 Broadening the Conversation

One of the proponents of these international solutions and broad conversations is Baylis (2017a and 2017b), who writes:

> "Why struggle? Because although all humans have a common interest in the human genome, much work is needed to identify other common interests that might rightfully guide policy deliberations. Hence, there is merit in [...] embracing the challenge of seeking broad societal consensus on this ethically controversial issue." (Baylis 2017b, 3)

Elsewhere, she continues defending the view that citizens should work towards broad societal consensus, because "[w]e are talking about nothing less than the future of the human species. No decisions about the modification of the germline should be made without broad societal consultation" (Baylis 2017a). Similarly, the statement produced after the International Summit on Human Gene Editing in December 2015 states that: "It would be irresponsible to proceed with any clinical use of germline editing unless and until [...] there is broad societal consensus about the appropriateness of the proposed application." (The National Academies 2015)

The problem with international and broad solutions is that we, members of *Homo sapiens*, often disagree. We disagree about everything: from climate change to national health services, from redistribution of income and wealth to genetically modified organisms, from vaccinations to sustainable diets. Our disagreements span from the evaluation of scientific methods to ethics and policy. They occur between and within countries, depending on personal and collective values, beliefs and practices, and – importantly – political and cultural frames (Jasanoff 2011). We tend to see the world through "different moral lenses" when it comes to values we hold dear (Greene 2014, 68).

In the face of all these disagreements, we can adopt different strategies, including throwing in the towel and not attempting to build such broad social consensus, or forcing the view held by the majority on the minority. Broad societal consensus might be unachievable and perhaps even undesirable. It may be unachievable due to our differing moral worldviews and our tribal tendencies to defend them (Greene 2014; Haidt 2012). It may also be undesirable inasmuch as it requires those participating in the debate on the ethics of genome editing to relinquish at least some of these moral views, or to restrict them to the private sphere and not bring them to the fore/um. It requires doing ethics, not inside the cave and among philosophers and ethics experts, but in the public square, where different moral views and "competing rationalities" coexist (Häyry 2010, 48). This is what commitment to a deliberative conception of democracy would encourage us to do (Cohen 2003; Gutmann and Thompson 2009). It would encourage us to engage in a process of public reasoning, where we collectively discuss contrasting moral views and engage in "visions of progress that are collectively defined, drawing on the full richness of democratic imagination" (Jasanoff et al. 2015). As we move from bioethics journals to institutional design and public policy, we need to be able to create an inclusive space for discussion of the ethical, political and social challenges raised by genome editing. We need to be able to develop policies that take account of these challenges and the competing moral views and values that underpin them.

For this reason, I side with Baylis' general intuition and with her plea for broad conversations on genome editing.[2] My view on the importance of an inclusive dialogue is motivated by a commitment to a deliberative understanding of democratic decision making, but also, as Peter Mills (2017) puts it, because we are in the field of "geo-ethics", where what matters are the "interactions of formal institutions, organisations and polities that recognise the potential for the spatiotemporal diffusion of genome editing technologies across political geographies and legal jurisdictions" (Mills 2017, 69). It is within and between such diverse and broad

2 I am not convinced by Baylis' idea of "broad societal consensus". I see the rhetorical appeal of such a proposal, but I think that aiming for "broad conversations" and trying to resolve, even partially, the disagreement is a) more feasible and b) more desirable. My argument then hinges more on the idea of "partially overcoming the disagreement", and trying to have "broad conversations" (i.e. as inclusive as possible), than on Baylis' idea of "broad societal consensus".

contexts, within and between "different social and cultural realities", that these technologies will be in play (Mills 2017).

What then should we do with genome editing? Should we take such competing moral views into account? Should we take all of them into consideration, or are there some views which should not be accepted in the public arena? Despite people's tendency to favour the moral views that resonate with their own and to discard opposing views as flawed, immoral, unacceptable etc., when it comes to developing policies, there are very good reasons to try to find certain overlaps among differing "moral tribes" (Greene 2014). I return to these questions (and to a tentative response to them) in section 3. Before that, I briefly explain how genome editing could be used in the context of assisted reproduction and then delineate some of the arguments put forward against and in favour of this possibility.

3 Reproductive Options and Genome Editing as a New Possibility

Couples who do not rely on assisted reproduction, and who wish to reduce the risk of transmitting a genetic mutation to their offspring, could refrain from having genetically related children and turn to adoption.[3] Alternatively, prospective parents can rely on reproductive technologies, such as IVF with gamete donations (third-party reproduction), or screening technologies, such as pre-implantation genetic diagnosis (PGD), to decrease the risk of having children with the genetic mutation they carry. In some cases, PGD is not an option as none of the embryos created through IVF is free from the undesirable genetic mutation (Vassena et al. 2016). For instance, when one of the prospective parents is homozygous for a dominant genetic disorder, the risk of transmission to offspring is as high as 100%, and hence no mutation-free embryos can be obtained. When prospective parents are both heterozygous for a dominant genetic disorder, the risk of transmission is as high as 75%, making the chances of finding mutation-free embryos significantly low. Another case where PGD is not effective is when both parents are homozygous for a recessive genetic disorder (the risk of transmission to offspring is as high as 100%), meaning that they both carry two variants of the disease-causing gene

3 I do not discuss this option in this chapter. Rather, I focus on the available options granted by existing and new assisted reproductive technologies. For a discussion of adoption, and why it is a morally preferable alternative to assisted reproduction, see for instance de Melo-Martín (2017a) and Rulli (2016).

(Nuffield Council on Bioethics 2016; Vassena et al. 2016). In such cases, genome editing could be employed directly on the embryos created with IVF to "correct" mutations which potentially lead to monogenic diseases, and to correct mitochondrial DNA mutations which lead to mitochondrial DNA diseases (Vassena et al. 2016).[4] Using genome editing on early embryos could give to prospective parents, who are at risk of transmitting genetic mutations to their offspring, a chance at having children who are genetically related to them, while decreasing the risk of transmitting the parents' genetic mutations.[5]

3.1 Genome Editing: Proponents and Critics

Research on human embryos with CRISPR technology is still in an early stage and only a few experiments have been carried out thus far (Vassena et al. 2016). Despite this, the possibility of allowing clinical research has been discussed recently (Chan et al. 2015; Gyngell et al. 2016; The National Academies of Sciences, Engineering and Medicine 2017; Reyes and Lanner 2017; Vassena et al. 2016). Critics advance two related, precautionary reasons against clinical applications of genome editing on human embryos or gamete cells: that these applications would lead to germline modifications and that assessing the safety of these applications would be either problematic or impossible. Those who worry about germline modifications view them akin to an (ethical and biological) "line that should not be crossed" (Collins 2015; UNESCO IBC 2017).[6] The worry about germline modifications expresses two underlying types of concern: deontological concerns about the intrinsic wrongness of introducing inheritable changes in the human genome, and consequentialist concerns about unforeseen effects for future generations, technology's running astray and the possible resurgence of eugenics (Lanphier and Urnov 2015; MacKellar 2017; Pollack 2015).

The first type of concern, which I refer to as "deontological", implies a principled position against germline modifications, regardless of the morality of the outcomes

---

4 Currently, the United Kingdom is the only country where mitochondrial DNA replacement techniques are allowed and regulated. Such techniques represent the only existing method for couples in which one member is affected by a mitochondrial condition to have genetically related children.

5 I do not discuss alternative forms of procreation, that do not entail genetic parenthood, in this chapter. I discuss the ethics of using new technologies to have genetically related, healthy babies and the broader societal implications of such decisions in Cavaliere 2017a.

6 For a discussion of this claim, see: Camporesi and Cavaliere 2016.

of such practices.[7] For instance, the 2017 statement of the UNESCO International Bioethics Committee (IBC) reads:

> "Interventions on the human genome should be admitted only for preventive, diagnostic or therapeutic reasons and without enacting modifications for descendants. The alternative would jeopardize the inherent and therefore equal dignity of all human beings and renew eugenics." (UNESCO IBC 2017)

Sometimes this principled objection echoes concerns regarding the intrinsic and inviolable value of human dignity and of what is "naturally" produced (Kass 1997; Sandel 2007), as opposed to what is technically designed.

The second type of concern, which I defined as "consequentialist" due to its focus on the state of affairs possibly effected by genome editing, encompasses objections related principally to the unforeseen effects of modifying the human germline, and the impossibility of thoroughly assessing the safety of these technologies (Baylis 2017b; Lanphier and Urnov 2015). At the current stage, safety is indeed an issue, and the efficiency of genome editing on embryos is still low, with mosaicism in edited embryos (i.e., edited embryos showed a mixture of edited and unedited cells) being the main known drawback of these technologies (Vassena et al. 2016).

3.2 The Case for Genome Editing

Those who argue in favour of the use of genome editing in the context of assisted reproduction agree that safety concerns must be thoroughly assessed before proceeding. Indeed, even the most vocal proponents of genome editing, and new reproductive technologies more generally, defend their clinical uses via the important caveat of assessing the risks that these technologies may pose to future children[8] (and the children of these children) (Chan et al. 2015; Harris 2016; Savulescu et al. 2015). For instance, Savulescu and colleagues (2015) argue that there is a moral imperative to continue gene editing research on human embryos in order to reduce off-target mutations and other safety issues, since "[a]dvanced and precise gene editing techniques could reduce the global burden of genetic disease and potentially benefit millions worldwide" (Savulescu et al. 2015, 476). Gyngell and colleagues (2016) focus

---

7 It also encompasses the objection to so-called enhancing uses of technologies, i.e., uses that are not aimed at treating genetic conditions, but at changing non-diseased traits (such as height, eye-colour or intelligence).

8 A position that represents a *de facto* argument in favour of gene editing research (Savulescu et al. 2015).

on the clinical application of genome editing (especially in the context of assisted reproduction) and state that there is a "significant medical case for pursuing GGE [germline gene editing]" to prevent the occurrence of genetic diseases in future generations (Gyngell et al. 2016, 499). Two sets of arguments have been advanced in favour of using genome editing technologies on embryos to correct genetic abnormalities, provided that safety concerns are properly addressed (Gyngell et al. 2016; Reyes and Lanner 2017). The first argument is grounded in the importance of procreation for individuals' well-being and for their flourishing. The argument of reproductive freedom states that people should be free to decide in matters of procreation and that no third-party should interfere with such freedom (if no harm is foreseen) (Harris 1992; Robertson 1994). Genome editing would enhance prospective parents' reproductive freedom as it would allow them to pursue their preferred life plan (i. e., having children) in the way that most resonates with their wishes and desires (i. e., having genetically related children who are free from the genetic mutation they themselves carry). Another argument in favour of genome editing focuses on the benefits of using this technology in the context of assisted reproduction for the offspring and for future generations in general. For instance, Harris (2015) argues that:

> "All of us need gene editing to be pursued, and if possible, made safe enough to use in humans. Not only to pave the way for procedures on adult tissues, but to keep open the possibility of using gene editing to protect embryos from susceptibility to major diseases and prevent other debilitating genetic conditions from being passed on through them to future generations." (Harris 2015)

Genome editing would allow for the editing of the genetic makeup of IVF embryos in vitro, who would consequently develop into mutation-free offspring. Genome editing would prevent the occurrence of genetic diseases both in the first generation offspring as well as in future generations, while PGD can sometimes only prevent the occurrence of genetic diseases in the child that develops from the implanted embryo (Gyngell et al. 2016).

4 How Do We Argue About New Technologies?

Let me take stock of what I have said so far. There are certain arguments against the use of genome editing (where germline modification and concerns with the safety of this technique are concerned) as there are arguments in favour of its use (where prospective parents' reproductive freedom and children's welfare are concerned).

What then? How are we going to find a path through what Greene (2014) describes as the "morass of competing moral values" (henceforth simply "the morass" or the "moral morass" [Grenne 2014, 185] and move towards a debate on new technologies that can constructively foster institutional design?

In this section, I focus on what I see as deep disagreement regarding both the moral standing of genome editing and the regulatory approaches that can be put in place. I explore different strategies that could be employed to overcome this deep disagreement, while indicating the foreseeable obstacles to the completion of this task. The first two options (and the related strategies) are what I see as the business of bioethics, or at least the business of many scholars working in bioethics. One option is to focus squarely on the disagreement by politely pointing out to those who hold a different moral view; Leveling the accusation to them that they are committing factual mistakes: They have failed to grasp how science and technology really work. The second option for dealing with existing disagreement is to try to persuade them of, or (at worst) to force onto them certain moral views by using sound philosophical arguments.

In my view, both options and the related strategies to overcome the deep disagreement are theoretically flawed and may result in states of affairs that do not promote our well-being. In the first two sections I explain why I believe these two strategies will not bring us through the morass. Then, in the last section, I provide some insights from moral psychology and argue that that literature can help us going forward to build a debate that can constructively aid the development of policies for the regulation of new technologies.

4.1 Option 1: It's All a Matter of Facts

One possible interpretation of the disagreement is to think about it in terms of a *misunderstanding of the relevant facts* of genome editing. This is an appealing interpretation as it could offer a potential strategy (the path through the morass that we are seeking) for overcoming the disagreement: getting the facts straight.

This interpretation is implicitly and, in certain instances, explicitly endorsed by those who criticise the arguments of other scholars for holding a different moral view concerning a new technology. Both groups often hint at the incapacity of "opponents" to understand how policy-making, scientific research, or the respective technology work. Consider the questions of uncertainty and safety. Both sides (those in favour of continuing genome editing research and those who oppose it) agree that, at present, it would be premature to use genome editing in the context of assisted reproduction and that the primary application of genome editing should be research-oriented. The problems (and the disagreements) arise when future applications are speculatively considered, and when the question is posed concerning when, if ever, genome editing applications on human embryos will become more successful (e. g., when they might have fewer off-target effects, or cease to give rise to mosaic embryos). For proponents of human genome editing the questions of uncertainty and safety are seen as contingent, rather than fundamental, impediments. For them, conducting experiments genome editing will lead to overcoming impediments such as mosaicism, to decrease the off-target effects and to increase the reliability of the technology. They focus on *when* (rather than *if ever*) genome editing will be safe enough to proceed. On the contrary, those who are sceptical about the benefits of the technology, or who oppose it in principle, consider questions of safety to be the end of the debate rather than the beginning, a reason not to carry out potentially dangerous research rather than a reason to further pursue technical knowledge. They see the impediments described above as evidence for current (and often future) unsafety of the technology.

So, what are the relevant, the right facts? And, especially, can scientists, policy-makers and ethicists settle once and for all the questions of uncertainty and safety? Spoiler alert: I argue that it is highly unlikely that they are able to do so. Other debates about whether new assisted reproductive technologies should be allowed can help to illustrate how safety-questions are rarely a matter of factual disagreement alone.

Two techniques for the replacement of maternal mitochondrial DNA in embryos created with IVF, using the genetic material of the prospective parents and the mitochondrial DNA of a donor, have been recently approved (2015) in the United Kingdom. The decision of the Parliament to approve these techniques was surrounded by a heated debate both within the United Kingdom and abroad. In a lively exchange between John Harris and Inmaculada de Melo-Martín, two philosophers who work in the field of bioethics, the question of safety was extensively discussed (Harris 2016; de Melo-Martín 2017b). Harris, a known proponent of MRTs and other assisted reproductive technologies, argued that the right question to ask was not whether MRTs were "safe", but whether they were "safe enough, given the balance of risks and benefits", adding that MRT "from a safety perspective was regarded as good to go" (Harris 2016, 10–11).[9] On this very same point (uncertainty and safety) de Melo-Martín responded: "Contrary to Harris, I believe that the evidence

9 Harris here refers to the multiple reviews of the "scientific methods to avoid mitochondrial diseases", carried out in the United Kingdom by the Human Fertilisation and the Embryology Authority (HFEA) in 2011, 2013, 2014, and 2016, prior to the approval of MRTs in the United Kingdom.

about the safety of these technologies is at this point completely inadequate." (de Melo-Martín 2017b, 161)

So, are MRTs safe or not? What are the facts of the matter? What does the evidence tell us? One possible answer is that either Harris or de Melo-Martín is simply wrong and has failed to understand correctly the scientific evidence regarding the safety of these techniques. Certainly, Harris thinks so about de Melo-Martín (and possibly vice versa), but this is precisely the point. Either one of them is right and the other is wrong (factually, that is), or there must be an alternative explanation for the disagreement.[10] And the very existence of such disagreements between two respectable scholars,[11] and especially within scientific, bioethical, and political communities, should set alarm bells ringing.

Safety is largely a contextual matter.[12] As Harris rightly points out, to say that a certain technology is "safe" hinges on how its risks and especially its benefits are evaluated – and how costs and benefits would be distributed. In other words, risks and benefits are contingent on the position of individuals in socio-economic, geopolitical, gender related and other aspects, and are very often unevenly distributed across the population (Prainsack and Buyx 2011).[13] Whether a technology is considered safe also depends, crucially, on whether approving the technology in question would violate principled positions in favour and against technological developments more generally (or principled positions regarding that specific

10 It is important to note that I believe that Harris is right in considering safety-questions in terms of "safe-enough" questions. When I say that one of them is right and the other is wrong I do not refer to their assessments of what safety means, but rather to questions regarding the benefits of MRTs. These questions, as I explain below, are in my view not entirely matters of fact.

11 I am not saying that respectable scholars are never factually wrong, but I am favouring an alternative interpretation to explain this specific disagreement.

12 Importantly, considering safety a contextual matter does not mean either that it is impossible to hold a realist position regarding science and technology, or that we should give up trying to assess the safety of new techniques. For a discussion (and defence) of objectivity and realism in scientific research see for instance Kitcher 2001.

13 Additionally, as I argue elsewhere, the ethical assessment of new techniques ought not only to rest on a cost/benefit analysis, but also on an evaluation of existing alternatives, including those that do not rely on biomedical means. In other words, whether genome editing really represents a worthy alternative to existing options (such as PGD) depends on the extent to which the welfare of the future child can be put at risk in order to allow couples to have a genetically related child. Regulators and ethicists that argue in favour of eventually replacing PGD with genome editing, and couples for whom PGD does not represent an option, will have to consider whether reproductive autonomy should trump questions about the welfare of the child, in light of uncertainty (Cavaliere 2017a).

technology). Principled positions, positions regarding what is morally right and wrong, being derived from higher moral principles,[14] influence our understanding of the science of genome editing and our assessment of the risks that the technology may pose to future generations. This does not mean that it is impossible to assess the safety, the risks or, more generally, the facts of scientific breakthroughs, nor that we should give up trying to do so. It also does not mean that we blindly follow high moral principles to the point of drawing conclusions that are independent of previous reflections, conversations, and our own intellectual work in general. It simply means that principled positions play a role in these assessments. While not denying that some empirical questions will be eventually put to rest, it is nevertheless important to note that a consensus on the question of safety will be hard to reach, due to the competing values at stake in stakeholders' assessments. Those who take a precautionary stance concerning technological development will favour existing technologies over newly discovered ones, while those generally in favour of technological development will be ready to accept a higher degree of risk in the name of such development and of the potential benefits it may yield for those in a position to enjoy such benefits (Prainsack and Buyx 2011).

In the past two decades, moral psychologists have shown how moral intuitions (i.e., intuitions regarding the rightness and the wrongness of a certain practice) play a role in the formation of moral judgments, and how the rational argument that follows is a *post hoc* rationalisation of the initial, intuitively formed, judgment (Greene et al. 2001; Haidt 2001; 2012). These principled positions against or in favour of a certain practice (or technology) are not derived after a thorough assessment of the risks, the benefits, the safety etc. of the practice, but intuitively and automatically, prior to the thorough assessment of the available evidence (Haidt 2012; Greene 2014). According Haidt (2001, 2012) and other moral psychologists (Greene et al. 2001), moral judgment is not the end point of a rational process in which, like zealous judges, people collect the available evidence (for example, regarding the safety of genome editing), thoroughly assess it, and only then come to a certain moral conclusion (say, a green or a red light for genome editing in the areas of basic research or assisted reproduction). When we argue about moral questions and moral beliefs people are more akin to lawyers who try to defend their original thesis (i.e. genome editing is safe enough or is not safe enough) by constructing a *post hoc* rational justification for it (Baumster and Newman 1994; Ditto et al. 2009; Haidt 2001, 2012). As Wright puts it:

14 When I say that these positions are derived from higher moral principles I do not mean that they are consciously derived from these principles, but that peoples' moral judgements are informed and derived by more general principles about right and wrong.

> "The brain is like a good lawyer: given any set of interests to defend, it sets about convincing the world of their moral and logical worth, regardless of whether they in fact have any of either. Like a lawyer, the human brain wants victory, not truth." (Wright 1994, 280)

The principled positions that we hold inform our assessment of safety, risks and benefits. They inform, in other words, the facts of the matter regarding a certain practice. It is not only that evidence in psychology shows that we are all very good in finding evidence that resonates with our principled moral position and with our beliefs in general. Studies focusing on the so-called "confirmation bias" (Shaw 1996), on "motivated reasoning", and especially on "motivated moral reasoning", show that we are much more likely to be sceptical and to question evidence that contradicts our principled position than evidence that supports it (for a review of the relevant social psychology literature, and a discussion of "motivated moral reasoning", see: Ditto et al. 2009). Motivated moral reasoning describes situations in which "Judgment is motivated by a desire to reach a particular moral conclusion." Thus, these are

> "situations in which an individual has an affective stake in perceiving a given act or person as either moral or immoral, and this preference alters reasoning processes in a way that adjusts moral assessments in line with the desired conclusion." (Ditto et al. 2009, 312)

Even if we have a conscious preference for objectivity in judgment, we often have unconscious directional motivations to reach certain conclusions that are preferred over others. This, again, does not mean that we are unable to reason ourselves away from such conclusions, nor that motivated moral reasoning is a robust and unmodifiable characteristic of the moral mind. It just means that we need to be aware of our own biases, preferences and moral views when we assess the evidence regarding, for instance, the safety of a new technology.

Harris, Savulescu, and others are known proponents of new technologies. De Melo-Martín, Baylis, and others who have argued against genome editing and MRTs, have held similar position with respect to other new technologies. Moreover, all of them are seasoned bioethicists capable of reading scientific papers correctly. Yet, for all of them, safety questions are either insurmountable or merely a contingent matter. So, who holds the truth of the matter concerning the safety of genome editing? The psychological literature on the role of moral intuitions in assessing the evidence, as well as the deep disagreement among bioethicists, scientists and policy-makers concerning safety questions, seem to provide at least a *prima facie* reason to be sceptical that to overcome the disagreement it would be sufficient to merely getting the facts straight. This makes the idea of grounding a broad consensus on facts alone a very risky project.

4.2 Option 2: May the Best (Argument) Win

Another strategy for overcoming the disagreement (which, I hope we have established, is at least also a moral disagreement) is to play the philosophers' beloved game and stage a good old rhetorical battle in which the best argument wins. Applied to the context of genome editing and assisted reproduction, with the (regrettable) impossibility of organising a global debate competition, the strategy would be to assess the arguments in favour of and against genome editing. This appears to be a good idea, and indeed almost the entire business of certain philosophical and bioethical traditions has been about doing exactly this. To assess the arguments in favour of and against the use of genome editing in assisted reproduction we could check for logical mistakes, for inconsistencies, for the correctness of the premises, and for the soundness of the conclusions. Yet, should we then also attempt to evaluate moral arguments on the basis of the state of affairs that the normative positions will bring about? Or should we focus on the arguments' resonance with a universal moral law? Or on what the arguments say about the moral character of the speaker? The appraisal of philosophical arguments, the different methods for conducting such an appraisal, and the conditions for an argument to be valid are disputed. What matters for the present discussion is that we be able to agree on certain minimal standards an argument should meet in order to be considered in the debate on genome editing. Nevertheless, I fear we would be left again with sound, logically valid, but still contradictory arguments.

Häyry (2010), for instance, talks about the different moral arguments, in favour of and against new genetic technologies, in terms of "competing rationalities".[15] These are divergent rational moralities that can be simultaneously held. According to Häyry, then, we should resist the temptation to compare such rationalities and elevate one of them above the others. Instead we should limit ourselves to checking for internal the consistency of the argument and their external consistency with how things are in the world (Häyry 2011). So, for instance, it is uncontroversial to reject as a fallacy the argument, "germline modifications are morally wrong because

15 Häyry (2011), defines rationality in "nonconfrontational" terms: "A decision is rational insofar as it is based on beliefs that form a coherent whole and are consistent with how things are in the world; and it is aimed at optimising the immediate or long-term impacts on entities that matter" (Häyry 2011, 43).

yesterday my grandma told me so" (arguments from authority should set off the alarm bells), or equally, arguments that are incoherent and inconsistent. However, it may be much harder to likewise reject arguments that defend positions like the following ones: "[G]ermline modifications are morally wrong because they could be harmful to the children born as a result of the application of CRISPR"; or "germline modifications are morally obligatory because they enhance parents' reproductive freedom, which is a fundamental moral right"[16].

Fallacious, illogical arguments can and should be criticised, especially in matters of science and ethics (Harris 2011). This is what participants in the debate on the ethics of genome editing (and other assisted reproductive technologies) do on an almost daily basis. There are two problems with this strategy, if our goal is not to find the best philosophical argument (i.e., "do moral philosophy"), but to build a space where coexisting moral positions are discussed and some degree of common ground is found (i.e., to "do public policies with insights from moral philosophy", and, I contend, moral psychology).[17] In these debates, the best argument – one that shows no logical mistakes, while exhibiting internal consistency, relying on valid premises and inferring valid conclusions – rarely *wins*. In other words, even if we agree on an independent strategy to establish who should win the prize,[18] and even if we pursue this strategy, we are left with a very good argument that may still fail to convince people who hold a different moral view (Camporesi and Maugeri 2011). We are left with, as Camporesi and Maugeri (2011) put it: "[A] cornucopia of ethical perspectives, each internally consistent but providing mere philosophical amusement." (Camporesi and Maugeri 2011, 255) This is a common problem in bioethics and in moral philosophy in general. An example: The argument that genome editing should be allowed, since it is in the interests of prospective parents to choose their

16 Personally, I am very sceptical of deontological arguments appealing to nature, sanctity, or rights. What I want to highlight here is that it is a very complex and perhaps unfeasible task to a) find an independent way to evaluate these arguments (i.e., an evaluative strategy that is not dependent on one's own moral and meta-ethical convictions), and b) to agree on who should decide which strategy is the most appropriate. For a thorough case against appealing to rights, see Greene 2014 (Chapter 11, Part V).

17 One might say that the business of bioethics is not to build a space for a mutual exchange regarding differing moral positions, but to let the "best" moral argument win. Besides the procedural problems briefly described above, what we are trying to achieve here (i.e., what I am discussing in this chapter) is broad societal consensus as well as finding at least some common ground to move forward in the debate on genome editing, and building a regulatory framework that helps to alleviate "the burden of human existence" (Harris 2016), whilst still accommodating opposing views as much as possible.

18 See note 16 above for an explanation of "independent strategy" and of the procedural problems that we may encounter.

preferred reproductive method, and to have their freedom in matters of procreation respected, is an argument that may be more intuitively appealing for someone who a) holds dear reproductive autonomy and freedom in general, and b) is sympathetic to the idea that reproductive freedom promotes well-being is an important value that should be respected. What I mean is that the pull of this argument may work on people who believe that freedom in matters of reproduction should trump other considerations, but not on those who, for instance, hold that assisted reproductive technologies are a mere commodity, and that research agendas should prioritise other, more pressing issues.

Besides these problems (of persisting moral disagreements and competing values), the problem with the "best argument" approach to the ethics of genome editing is that, even if it was possible to overcome the procedural challenges in selecting it, we may still not win as a community and as individuals. In other words, what we are faced with is not a philosophical problem, but a political problem. We may have very good arguments, sound and valid philosophical arguments, which still fail to promote our well-being and our happiness (broadly conceived, that is: independently from one's own conception of well-being), which is arguably what these technologies, and perhaps medicine in general, should promote.

4.3 Perhaps not Everything is Lost. The "Listening Mode" and the "Meta-Morality"

If my arguments (and some of the evidence from moral psychology) have managed to convince you, then at this point you should be at least a little sceptical of two options for overcoming the disagreement concerning the use of genome editing in assisted reproduction: explaining the facts of the matter to those who still fail to understand them, and letting the best argument win. Both strategies are, in my view, theoretically flawed and may end up bringing about state of affairs that do not promote our well-being. The deep moral disagreement surrounding genome editing technologies and their applications seems to be an obstacle we cannot ignore, if we want to move forward (by either banning or regulating applications of these technologies). What then should we do?

A possible diagnosis of the deep moral disagreement we experience in our everyday lives is that we have different moral intuitions and we hold different moral beliefs (Greene 2014; Haidt 2012). Religious scholars, feminist scholars, liberal philosophers, critical theorists, and lay people who find themselves embedded in differing political, socio-economic and cultural context, react in differing ways to contemporary (moral) controversies. Gay marriage, redistribution of income and wealth, but also debates about abortion and genome editing, stir controversy because we hold different moral views, some of which are *post hoc* rationalisation of differing moral intuitions. Moral psychologists Jonathan Haidt and Joshua Greene offer different normative prescriptions for overcoming the disagreements and learning to cooperate between different moral tribes. In this final section of my chapter I briefly outline their normative prescriptions. My conclusion is that they are both illuminating for the way we reflect on genome editing and the way we develop strategies to regulate new technologies beyond national borders.

Haidt's social intuitionist model (Haidt 2001; 2012) shows that people are good at finding evidence that confirms their initial moral judgment (remember the lawyer analogy!). What can make us change our minds about our moral beliefs is the multiplicity of occasions where we find ourselves challenged by others (what Haidt calls "reasoned persuasion" [Haidt and Björklund 2008]), and by the social environment in which we are embedded (what Haidt refers to as "social persuasion"[19] [Haidt and Björklund 2008]). Moral reasoning needs to be understood in terms of a social process, in which people embark alone on the search for evidence and come together for the appraisal of such evidence. We need to be aware that we are self-righteous creatures, with a tendency to form moral beliefs from our intuitions and to rationalise them afterwards in our exchanges with others. To better get along with each other, and to find a path through the moral morass of genome editing, we need to acknowledge that our disagreements do not necessarily and solely stem from factual mistakes, wrong arguments, and wrong moral beliefs,[20] but from our competing rationalities and differing intuitions. Haidt's central normative prescription is to be more open towards those who hold views different from ours, to be aware that we are all biased by our moral intuitions (all of us, not only those who disagree with us), and to be aware that some of our grand arguments may simply be *post hoc* rationalisations of gut feelings. Thus, the Haidtian path through the moral morass of genome editing encourages us to abandon the mode of combat, and put ourselves in the listening mode, when we discuss the ethics of genome editing, its applications, and the regulatory frameworks which should be adopted.

Greene (2014) accepts Haidt's premises, shares his evolutionary understanding of morality, his view on the role of intuitions in the formation of (only some types

19 Social persuasion refers to the fact that persuasion does not only happen when others give us reasons, arguments or new evidence, but also simply by virtue of our being "sensitive to and influenced by what other people think and feel" (Haidt Haidt and Björklund 2008, 192).

20 By wrong moral beliefs, I mean the moral beliefs that we do not share and that we may have failed to understand.

of) moral judgments, and importantly, his (above outlined) normative prescription. However, despite this broad agreement, he argues that Haidt's normative prescription alone will not bring us much further in our moral disagreements. In other words, listening and being open to opposing views are all good strategies as far as they go, but they are unlikely to bring us forward, if broad societal consensus[21] is what we aim for.

Over the last 20 years, Greene has examined the brain scans of people while they were reflecting on moral dilemmas such, as the trolley problem.[22] Greene observed that different areas of the brain (the ventromedial prefrontal cortex, VMPFC, and the dorsolateral prefrontal cortex, DLPFC) showed more activity depending on which variation of the trolley problem was proposed. The footbridge case elicited more brain activity in people's DLPFC, the brain area associated with more logical, calculating cognitive capacities. On the contrary, more impersonal dilemmas such as the switch case elicited more brain activity in the VMPFC and in the amygdala, the brain regions associated with emotions. The only difference between the footbridge case and the switch case is whether the man scarified to save five is directly pushed down from the bridge to the track (and hence killed to prevent the trolley to kill the other five people) or he is killed by hitting a switch that diverts the trolley from a track where five people are standing to the track where he is standing. In a series of experiments conducted by Greene and others they verified over and over how brain activity in the DLPFC was linked to choosing utilitarian solutions (i. e. killing one to save five), whilst activity in the VMPFC was linked to dilemmas that

21 Greene does not aim at broad societal consensus alone. He has a normative theory in mind for how that broad consensus ought to look. However, a discussion of his proposal is beyond the scope of this chapter.

22 The trolley problem describes a cluster of moral dilemmas that are all characterised by the choice of sacrificing one person in order to save five. The trolley problem was first discussed by Philippa Foot (1967) in an article discussing abortion and the doctrine of double effect. Different version of the problem and an analysis of the moral views that underpin them were later discussed by Judith Jarvis Thompson (1976). The two variations I am referring to here are the "switch case" and the "footbridge case". The trolley problem describes a situation in which a runaway trolley is barrelling down one track where there are five people tied up and unable to move. On another track there is one person, also tied up to the track and unable to move. In the switch case, a bystander could pull a lever and divert the trolley onto the track where only one person is tied up. In the footbridge case, the trolley is still barrelling down a track where five people are tied up and unable to move, but the only way the bystander can stop the trolley and prevent their death is by pushing a fat man (or a man with a large backpack) off a bridge. In both cases, the choice is between killing one person to save five or do nothing and let these five people die.

involved directly pushing and hence killing a person to save the other five. This led Greene to formulate the theory of the dual-process brain (or of the two moral minds), delineating the process that works in automatic-mode and it is guided by our unconscious, automatic, emotional responses, and the process that works in manual-mode and has "conscious access to the operative decision rule. The rule that maps the relevant features of the situations onto a suitable behaviour" (Greene 2014, 136). According to Greene's dual-process theory: "Reasoning frees us from the tyranny of our immediate impulses by allowing us to serve values that are not automatically activated by what's in front of us" (Greene 2014, 137). We still need the input of emotions for decision-making, for evaluating risk (Damasio 1994), and for avoiding the conflicts that hinder cooperation among members of the same moral tribe, but we also need more than this (Greene 2014).

Greene's normative prescription is that we need something more than emotions, than the automatic mode, to avoid conflicts between tribes that are moral in differing ways. Listening, avoiding self-righteousness and being more open for compromise are good steps. However, they are only first steps. We need all this and more to find a path through the moral morass. We need to work in manual mode to develop what Greene calls a "meta-morality":

> "A global moral philosophy that can adjudicate among competing interests of its members. A meta-morality job is to make trade-offs among competing tribal values, and making trade-offs requires a common currency, a unified system for weighing values." (Greene 2014, 15)

The manual mode, the one that makes us choose to kill one person to save five in both the variations of the trolley problem, is the mode that causes us to reflectively endorse the greater good. The mode that overrides the emotional rejection of killing an innocent person in the name of the greater good (saving five people) is utilitarian morality (which Greene calls "deep-pragmatism"). Greene's view that utilitarianism is a 'splendid idea', one we should all endorse, or at least strive to endorse in order to overcome our tribal conflicts, is an interesting view. It is also very controversial and has been criticised for a number of reasons, including the fallacy of deriving normative conclusions from factual claims (from the brain scans to deep-pragmatism as the morality that we ought to embrace), and critiques that are normally put forward against utilitarianism in general.

Despite the criticisms and despite detractors of utilitarianism, I think there is a lesson to be learned from Greene's proposal. What we need is to develop a meta-morality that causes us to transcend our tribal disagreements and that allows us to discuss together "what matters" (Parfit 2011). The Greenian path through the moral morass of genome editing encourages us to find a common moral language, shared values and shared moral beliefs as a starting point for discussing the ethics and policy questions of genome editing.

5 Conclusion

A decision about allowing genome editing in the context of assisted reproduction has to rest not only on a thorough assessment of the safety of the techniques, and their possible or likely benefits, but also on a democratic process that takes into account differing views and values (Cavaliere 2017b; Jasanoff et al. 2015; Kitcher 2001). In this chapter, I have focused on genome editing in the context of assisted reproduction, and on the calls for broad societal consensus in tackling this question. I presented two main arguments against this possibility: namely, safety and that these applications would lead to germline modifications. I then turned to arguments in favour of genome editing, such as the welfare of future children and peoples' reproductive autonomy. As many have argued, it may be too soon for a conclusive assessment of this possibility, if only for the dearth of empirical data regarding its safety and feasibility. Thus, what this chapter offers a basis to begin a discussion on the ethics of genome editing that is informed by the literature in moral psychology. I argued that we should focus on finding a common currency and shared moral values, a meta-morality that goes beyond the deep moral disagreements among us, and that allows us to speak a common language that enables a minimum of agreement among us.

References

Baltimore, David et al. 2015. A Prudent Path Forward for Genomic Engineering and Germline Gene Modification. *Science* 348: 36–38.

Baumeister, Roy F., and Leonard S. Newman. 1994. Self-Regulation of Cognitive Inference and Decision Processes. *Personality and Social Psychology Bulletin* 20: 3–19.

Baylis, Françoise E. 2017a. Human Genome Editing: We Should All Have a Say. The Conversation. https://theconversation.com/human-genome-editing-we-should-all-have-a-say-81797. Accessed 12 October 2017.

Baylis, Françoise. 2017b. Human Germline, Genome Editing and Broad Societal Consensus. *Nature Human Behaviour.* doi:10.1038/s41562-017-0103.

Brokowski, Carolyn, et al. 2015. Cutting Eugenics out of CRISPR-Cas9. *Ethics in Biology, Engineering and Medicine: An International Journal* 6: 263–279.

Camporesi, Silvia, and Giulia Cavaliere. 2016. Emerging Ethical Perspectives in the Clustered Regularly Interspaced Short Palindromic Repeats Genome-Editing Debate. *Personalized Medicine* 13: 575–586.

Camporesi, Silvia, and Paolo Maugeri. 2011. Genetic Enhancement in Sports: The Role of Reason and Private Rationalities in the Public Arena. *Cambridge Quarterly of Healthcare Ethics* 20: 248–257.

Cavaliere, Giulia. 2017a. Genome Editing and Assisted Reproduction: Curing Embryos, Society or Prospective Parents? *Medicine, Health Care and Philosophy*: 1–11.

Cavaliere, Giulia. 2017b. A 14-Day Limit for Bioethics: The Debate over Human Embryo Research. *BMC Medical Ethics* 18: 38.

Chan, Sarah, et al. 2015. Genome Editing Technologies and Human Germline Genetic Modification: The Hinxton Group Consensus Statement. *The American Journal of Bioethics* 15: 42–47.

Cohen, Joshua. 2003. Deliberation and Democratic Legitimacy. In *Debates in Contemporary Political Philosophy. An Anthology*, Derek, Matravers, and Jonathan Pike (eds.), 342–360. London: Routledge.

Collins, Francis S. 2015. Statement on the NIH Funding of Research Using Gene-Editing Technologies in Human Embryos. National Institute of Health (NIH). http://www.nih.gov/about-nih/who-we-are/nih-director/statements/statement-nih-funding-research-using-gene-editing-technologies-human-embryos. Accessed 12 October 2017.

Damasio, Antonio. 1994. *Descartes' Error: Emotion, Reason and the Human Brain*. New York: G.P. Putnam's Sons.

Ditto, Peter H., et al. 2009. Motivated Moral Reasoning. *Psychology of learning and motivation* 50: 307–338.

Foot, Philippa. 1967. The Problem of Abortion and the Doctrine of Double Effect. *Oxford Review* 5: 5–15.

Greene, Joshua D. 2014. *Moral tribes. Emotions, Reason and the Gap Between Us and Them*. London: Atlantic Books.

Greene, Joshua D., et al. 2011. An fMRI Investigation of Emotional Engagement in Moral Judgment. *Science* 293: 2105–2108.

Gutmann, Amy, and Dennis Thompson. 2009. *Why Deliberative Democracy?* Princeton: Princeton University Press.

Gyngell, Christopher, et al. 2016. The Ethics of Germline Gene Editing. *Journal of Applied Philosophy* 34: 1–16.

Haidt, Jonathan. 2001. The Emotional Dog and its Rational Tail: A Social Intuitionist Approach to Moral Judgment. *Psychological review*. 108: 814–834.

Haidt, Jonathan. 2012. *The Righteous Mind: Why Good People Are Divided by Politics and Religion*. New York: Pantheon.

Haidt, Jonathan, and Fredrik Björklund. 2007. Social Intuitionists Answer Six Questions about Morality. In *Moral psychology. Volume. 2: The Cognitive Science of Morality: Intuition and Diversity*, Walter Sinnott-Armstrong (ed.), 181–217. Cambridge: MIT Press.

Harris, John. 1992. *Wonderwoman & Superman: Ethics & Human Biotechnology*. Oxford: Oxford University Press.

Harris, John. 2011. The Challenge of Nonconfrontational Ethics. *Cambridge Quarterly of Healthcare Ethics* 20: 204–215.

Harris, John. 2015. Why Human Gene Editing Must not be Stopped. *The Guardian*, December 2.

Harris, John. 2016. Germline Modification and the Burden of Human Existence. *Cambridge Quarterly of Healthcare Ethics* 25: 6–18.
Häyry, Matti. 2010. *Rationality and the Genetic Challenge: Making People Better?* Cambridge: Cambridge University Press.
Jasanoff, Sheila. 2011. *Designs on Nature: Science and Democracy in Europe and the United States*. Princeton: Princeton University Press.
Jasanoff, Sheila, et al. 2015. CRISPR Democracy: Gene Editing and the Need for Inclusive Deliberation. *Issues in Science and Technology* 32: 37–49.
Kass, Leon R. 1997. The Wisdom of Repugnance: Why We Should Ban the Cloning of Humans. *Valparaiso University Law Review* 32: 679–704.
Kitcher, Philip. 2001. *Science, Truth and Democracy*. Oxford/New York: Oxford University Press.
Lander, Eric S. 2015. Brave New Genome. *New England Journal of Medicine* 373: 5–8.
Lanphier, Edward, et al. 2015. Don't Edit the Human Germ Line. *Nature* 519: 410–411.
MacKellar, Calum. 2017. The Gene Editing of Human Embryos and the New Eugenics. *Bioethica Forum* 10.
de Melo-Martín, Inmaculada. 2017a. *Rethinking Reprogenetics. Enhancing Ethical Analyses of Reprogenetics Technologies*. New York: Oxford University Press.
de Melo-Martín, Inmaculada. 2017b. When the Milk of Human Kindness Becomes a Luxury (and Untested) Good. *Cambridge Quarterly of Healthcare Ethics* 26: 159–165.
Mills, Peter. 2017. Lame Ducks Might Fly: Genome Editing, Global Consensus and Geo-Ethics. *Bioethica Forum* 10.
Nuffield Council on Bioethics. 2016. *Genome Editing. An Ethical Review*. London: Nuffield Council on Bioethics.
Parens, Erik, and Adrienne Asch. 2003. Disability Rights Critique of Prenatal Genetic Testing: Reflections and Recommendations. *Mental Retardation and Developmental Disabilities Research Reviews* 9: 40–47.
Parfit, Derek. 2011. *On What Matters. Volume One*. Oxford: Oxford University Press.
Prainsack, Barbara, and Alena M. Buyx. 2011. *Solidarity: Reflections on an Emerging Concept in Bioethics*. London: Nuffield Council on Bioethics.
Reyes, Alvaro Plaza, and Fredrik Lanner. 2017. Towards a CRISPR View of Early Human Development: Applications, Limitations and Ethical Concerns of Genome Editing in Human Embryos. *Development* 144: 3–7.
Roberts, Dorothy E. 1997. *Killing the Black Body: Race, Reproduction, and the Meaning of Liberty*. New York: Vintage Books.
Robertson, John A. 1994. *Children of Choice: Freedom and the New Reproductive Technologies*. Princeton: Princeton University Press.
Rulli, Tina. 2016. Preferring a Genetically-Related Child. *Journal of Moral Philosophy* 13: 669–698.
Sandel, Michael J. 2007. *The Case Against Perfection*. Cambridge: Harvard University Press.
Savulescu, Julian, et al. 2015. The Moral Imperative to Continue Gene Editing Research on Human Embryos. *Protein & Cell* 6: 476–479.
Shaw, Victoria F. 1996. The Cognitive Processes in Informal Reasoning. *Thinking & Reasoning* 2: 51–80.
The National Academies. 2015. On Human Gene Editing: International Summit Statement. http://www8.nationalacademies.org/onpinews/newsitem.aspx?RecordID=12032015a. Accessed 18 October 2017.

The National Academies of Sciences, Engineering, Medicine. 2017. *Human Genome Editing: Science, Ethics, and Governance*. Washington, DC: The National Academies Press.
Thomson, Judith J. 1976. Killing, Letting Die, and the Trolley Problem. *The Monist* 59: 204–217.
UNESCO International Bioethics Committee (IBC). 2017. UNESCO Panel of Experts Calls for Ban on "Editing" of Human DNA to Avoid Unethical Tampering with Hereditary Traits. http://en.unesco.org/news/unesco-panel-experts-calls-ban-editing-human-dna-avoid-unethical-tampering-hereditary-traits. Accessed 12 October 2017.
Vassena, Rita, et al. 2016. Genome Engineering Through CRISPR/Cas9 Technology in the Human Germline and Pluripotent Stem Cells. *Human Reproduction Update* 22: 411–419.
Wright, Robert. 1994. *The Moral Animal*. New York: Pantheon.

Kolibri 360 GmbH
Ferdinand-Jühlke-Straße 7
99095 Erfurt, Deutschland
produktsicherheit@kolibri360.de